BIOLOGICAL APPLICATIONS OF LIQUID SCINTILLATION COUNTING

BIOLOGICAL APPLICATIONS OF LIQUID SCINTILLATION COUNTING

Yutaka Kobayashi

The Worcester Foundation for
Experimental Biology, Inc.
Shrewsbury, Massachusetts

David V. Maudsley

The Worcester Foundation for
Experimental Biology, Inc.
Shrewsbury, Massachusetts

1974

ACADEMIC PRESS New York San Francisco London

A Subsidiary of Harcourt Brace Jovanovich, Publishers

ACADEMIC PRESS, INC.
111 Fifth Avenue, New York, New York 10003

United Kingdom Edition published by
ACADEMIC PRESS, INC. (LONDON) LTD.
24/28 Oval Road, London NW1

Library of Congress Cataloging in Publication Data

Kobayashi, Yutaka, Date
Biological applications of liquid scintillation counting.

Based on lectures given at the Worcester Foundation for Experimental Biology.
Includes bibliographical references.
1. Liquid scintillators–Congresses. I. Maudsley, David V., joint author. II. Worcester Foundation for Experimental Biology. III. Title. [DNLM: 1. Radiobiology. 2. Scintillation counters. WN650 K75b]
QH324.9.L53K6 574'.028 74-5693
ISBN 0–12–417250–4

PRINTED IN THE UNITED STATES OF AMERICA

Contents

3 GENERAL CONSIDERATIONS OF SAMPLE PREPARATION

4 TESTING AND SETTING UP A NEW COUNTER

5 BASIC PRINCIPLES OF THE APPLICATION OF RADIOTRACERS

Preface

The subject of radiation and related topics have carried to the public the specter of nuclear confrontation between nations and all associated horrors of an atomic holocaust. Progress in science, however, constitutes a double-edged potential, and as awesome as nuclear radiation is as an instrument of war it is also a powerful aid in the diagnosis and treatment of disease and in furthering our understanding of biological systems. Indeed, it is difficult to imagine a modern biological research program which does not, at some stage, utilize radioisotope techniques.

There have been several motives in writing this book. It collates a series of lectures given over a period of many years to the postdoctoral trainees in the various specialized training programs held at the Worcester Foundation for Experimental Biology. This institute has played a prominent role in the application and development of radiotracer techniques for the study of steroid hormones, and innovations and improvements in liquid scintillation counting have been an important element in the progress made. This book aims to fulfill some of the gaps left by the demise of the national training programs and the associated loss of formal course work. An extention of this objective has been to provide the basis for a training course in liquid scintillation counting appropriate for the university level. It is our experience that even recent science graduates have rarely seen or handled a liquid scintillation counter even though they may have taken a course in radiochemistry. Many teaching institutes persist in teaching radiochemical techniques using methods that are archaic, such as Geiger counting, and instruments that are obsolete. The universities could render the scientific research community a major service by altering the emphasis of their radiochemical courses and assuming the responsibility for teaching both practical and theoretical aspects of liquid scintillation counting.

The automated scintillation counter generates an enormous amount of data, and the experimental scientist is more concerned with its interpretation than establishing its validity. The stimulus to the imagination which results from observing a difference in count rate between two samples is not easily quenched by suggestions that it may simply reflect variations in counting efficiency. Scintillation counting is being used by an increasing number of investigators who have only a vague idea, for example, of the legitimacy of counting heterogeneous samples, the problems of determining counting efficiency in emulsion systems, or the relevance of instrument drift in low level double-isotope analysis. The consequences of this trend will be a further proliferation of the nonreproducible experiments which already constitute such a significant proportion of the scientific literature and a further diminution of the standards of technical excellence which are obligatory for sustained scientific progress. The problem, in part, is due to a lack of adequate training, but it may also reflect the present climate of biological research which tends to produce perishable publications without giving the same attention to the rigorous proof of observation that so often characterized the work of our scientific predecessors.

There is no defense for the improper use of a scintillation counter, and although it represents the "state of the art" in electronics one need not be an electronics engineer to operate it correctly. It does, however, require an understanding of the principles of the scintillation process, some knowledge of how the scintillation spectrometer operates, and an awareness of the requirements for proper sample preparation. It is our hope that this work presents the fundamentals of liquid scintillation counting in a manner which is simple, practical, and, above all, useful.

We would like to thank Walter Zydlewski who spent many hours explaining the electronics involved in the various components of the liquid scintillation spectrometer, Edward Polic who made many valuable suggestions and comments during the early writing of the text, and Richard Frank and Barton Laney for reviewing the manuscript. Finally, we are much indebted to Mrs. Virginia Barber, for without her tact, patience, and talent we would never have achieved our goal.

Yutaka Kobayashi
David V. Maudsley

1

Introduction

As late as 1939 it was generally considered that the existence or production of long-lived radioisotopes of biological interest, such as carbon-14 and tritium, was unlikely. With the development of a technique that permitted more intensive bombardment of specimens within a cyclotron, however, Ruben and Kamen (1, 2) produced carbon-14 by bombarding graphite with deuterons and thus initiated the radioactive era of biological research. The attributes of a long half-life of over 5000 years; the availability of a convenient method of assay, namely, gas ionization techniques; and the relative ease with which this radioisotope could be handled safely contributed to its rapid acceptance and use. The availability of carbon-14 has played and continues to play, a major role in advancing the state of biological knowledge. What is not so well known is the fact that tritium was actually produced shortly before carbon-14 (3) but the difficulty in assaying accurately the low-energy beta particle emitted by this isotope discouraged its application to biological problems. Although the potential utility of tritium as a biological research tool was certainly as great as that for carbon-14, the lack of a convenient assay method prevented its realization. The major isotopes used in metabolism studies prior to 1954 were therefore carbon-14, sulfur-35, and phosphorus-32.

The first commercial liquid scintillation counter was introduced in 1954 (4). Compared to the prevailing methods for counting isotopes, such as Geiger or proportional counting of solid samples or the tedious but accurate combustion method of gas ionization counting, the introduction of liquid scintillation counting offered many desirable features. These included high counting efficiencies for the soft beta emitters, such as tritium, carbon-14, and sulfur-35; ease of sample preparation; and a lack of self-absorption. The potential of the technique was immediately evident but the deterrent that initially limited its widespread use was the relatively high cost of the counter itself.

The major attraction of liquid scintillation counting was that it was the only convenient method available for assaying tritium. At the same time rapid advances were being made in preparative techniques that made possible the labeling of most natural products by tritium exchange (5). In economic terms, carbon-14 as barium carbonate is 10,000 times as expensive as tritium gas for equivalent amounts of radioactivity. Because tritium can be obtained in carrier-free form, with a specific activity of 57 Ci/mmole, the synthesis of tritiated compounds with very high specific activity is possible. A further consideration is that the inherent sensitivity of the instrument permits the use of smaller quantities of isotope than had previously been possible, thus effecting a saving on the cost of labeled compounds, and since 1960 liquid scintillation counting has been established as the dominant method for analyzing beta emitters commonly used in the bioscience laboratories.

REFERENCES

1. Ruben, S., and Kamen, M. D. *Phys. Rev.* **57,** 549 (1940).
2. Kamen, M. D., and Ruben, S. *Phys. Rev.* **58,** 194 (1940).
3. Alvarez, L., and Cornog, R. *Phys. Rev.* **56,** 613 (1939).
4. Rapkin, E. *In* "The Current Status of Liquid Scintillation Counting" (E. D. Bransome, Jr., ed.), p. 45. Grune & Stratton, New York, 1970.
5. Wilzbach, K. E. *J. Amer. Chem. Soc.* **79,** 1013 (1957).

2

Principles of Liquid Scintillation Counting

SCINTILLATORS AND THE SCINTILLATION PROCESS

Scintillation counting, dating back to W. Crookes in 1903, was one of the earliest methods known for measuring radiation (1). When a source of alpha radiation was to be measured, the early workers would place the source in front of a zinc sulfide screen in a dark room. The observer would see small flashes of light or scintillations appear on the screen and he would record by hand the number of scintillations observed. Crookes used this phenomenon as the basis for an instrument called a "spinthariscope," which consisted of a small tube with a lens at one end and a zinc sulfide screen at the other, with the alpha source placed between the two but nearer to the screen. Observation of the screen through the lens in a darkened room revealed flashes of light on various parts of the screen. By calibrating the area of zinc sulfide exposed to a weighed amount of an alpha source placed at a fixed distance from the screen, E. Regener (1908) was able to quantitate the production of alpha particles (2). It might be said therefore, that one of the original radiation detectors was zinc sulfide, the first scaler the human eye, and the first recorder the human hand. The obvious limitations imposed by this system and the development of radiation counting methods based on gas ionization led to the demise of visual scintillation counting in the early 1930's.

The development of liquid scintillation counting has been stimulated by two major events. The first was the discovery of organic compounds called "scintillators" and the other was the development of photomultiplier tubes. Scintillators have the property of absorbing radiant energy either in the solid state or in solution. The absorption of this energy by the scintillator results in the formation of excited atoms or molecules that then return rapidly to the normal or ground state, releasing as photons (light energy) and heat this excess energy derived originally from nuclear radiation. These scintillators are transparent to their emitted light, which is in the ultraviolet or visible range. The number of photons emitted is linearly related, approximately, to the radiant energy absorbed, and this fundamental property of scintillators allows the scintillation counter to be used not only as a counter but as a means of differentiating beta-emitting radioisotopes having different energies. For example, 1 μCi of carbon-14 yields about ten times the number of photons as 1 μCi of tritium, although they have identical disintegration rates of 2.2×10^6 per minute. Provided the two isotopes yield beta particles with energies sufficiently different from each other, both can be determined simultaneously. This is one of the most useful features of a liquid scintillation counter.

The term "liquid scintillation counting" is used because these scintillators are usually dissolved in a suitable solvent containing the radioactive material to be assayed. The photon production from this solution results from the following sequence of events: The energy of the beta particle emitted from a radioactive source is first absorbed by the solvent molecules, causing them to become excited. This excitation energy is propagated within the solvent and transferred to the scintillator (solute), causing the scintillator molecules to become excited, and when they return to their ground state they emit photons (light). The exact mechanisms of these energy transfers are not fully understood (3–6) but the following generalizations can be made: The initial transfer of energy from the beta particle to the solvent molecules is the most inefficient of the energy transfer steps of the liquid scintillation process leading to photon emission. This is deduced from the observation that the efficiency of energy transfer from the solvent, xylene, excited by three different methods to a scintillator, diphenyloxazole (PPO), at a concentration of 2 g/liter, is in excess of 90% (7, 8). In a comparable study, the efficiency of energy transfer from a primary scintillator, *p*-terphenyl, to a secondary scintillator, 1,1′,4,4′-tetraphenylbutadiene (TPB), dissolved in toluene has also been shown to be between 90 and 95% at the concentrations normally used for secondary scintillators (9). The short-range characteristics of weak beta particles, such as those of tritium (about 8 μm in water) and carbon-14 (less than 3 mm in water) ensures that all the energy of the emitted beta particles is

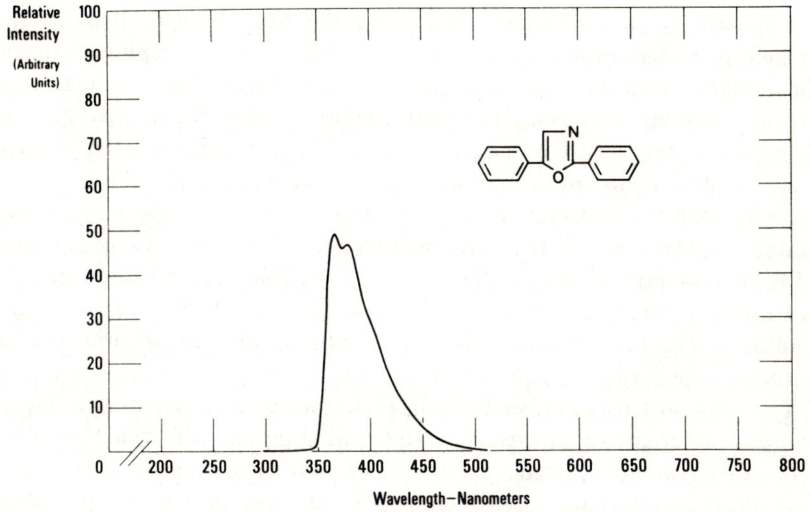

Fig. 1 Fluorescence spectrum of 2,5-diphenyloxazole (PPO) in toluene.

dissipated within the solution and absorbed by the solvent molecules. Only a small percentage (roughly 5–10%) of the total absorbed energy results in photon production (3, 6) and because the transfer of energy from the solvent to the scintillator is very efficient the major loss of energy as heat must occur during the first step. The scintillation decay time for a typical scintillator, such as 2,5-diphenyloxazole (PPO), in toluene is less than 3 nseconds (10). The fluorescence spectrum for this compound is shown in Fig. 1. The fluorescence maximum occurs at 365 nm, which is in the blue region of the light spectrum. This feature determines the spectral sensitivity required of the fluorescence detector used in liquid scintillation counting.

PHOTOMULTIPLIER TUBES

The second major event that encouraged the development of liquid scintillation counters was the introduction of very sensitive photomultiplier tubes. The photomultiplier is a vacuum tube that converts photons into electrical energy and consists of a photocathode and a series of dynodes. The photocathode and dynodes are coated with cesium–antimony, silver–magnesium–cesium, or similar alloys that have the property of emitting photoelectrons when exposed to photons. These photoelectrons are then accelerated by a positive potential into the first dynode. Because the dynode is coated with the same or similar material as the photocathode, the impact of each accelerated photoelectron causes the production of several secondary electrons. The electrons produced at the first dynode are then accelerated toward a second dynode. This process is repeated through a series of dynodes until the secondary electrons emerging at the last dynode reach the collector (anode) as an avalanche, and result in a measurable electrical pulse. One photoelectron may generate as many as a million secondary electrons. The overall electron multiplication can be controlled by changing the positive potential gradient (high voltage) applied to the tube. Again, this is a near-linear process, so proportional counting is possible.

The two most common varieties of photomultiplier tubes used in liquid scintillation counting are the box and grid type and the venetian blind type. A diagramatic box and grid type with four dynodes is shown in Fig. 2 to illustrate the operation of a photomultiplier tube. In this example, three photons impinge on the photocathode of the photomultiplier tube, resulting in the production of one photoelectron. The ratio of the number of photoelectrons released to the number of impinging photons expressed as a percentage is defined as the "photocathode quantum efficiency," which, in this case, is 33%. The single photoelectron produced by these photons is accelerated toward the first dynode by a positive potential applied at the dynode. On impact with the dynode, ten secondary electrons are produced, which are then accelerated toward the second dynode (which is more positive than the first dynode) and the multiplication process is repeated.

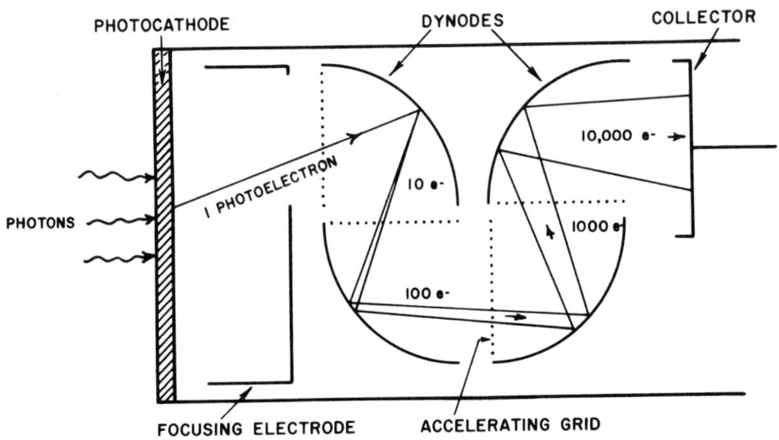

Fig. 2 A hypothetical photomultiplier tube using box and grid-type construction with four dynodes.

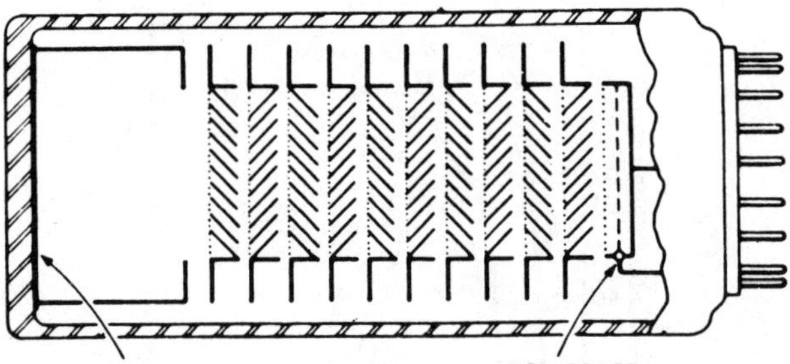

Fig. 3 A photomultiplier tube using venetian blind-type construction. (Courtesy of EMI.)

The grid, which is at the same potential as the dynode to which it is connected, prevents the electrons from falling back to their place of origin (previous dynode). The multiplication factor for each dynode is the total number of photoelectrons produced per incident electron which, in our example, is ten, giving an overall multiplication factor of 10^4 or 10,000 for this tube. The multiplication factor for a typical photomultiplier tube used in liquid scintillation counting can be as high as 50 million in a 13-dynode photomultiplier tube. In Fig. 3 is shown the essential construction of a ten-stage venetian blind type of photomultiplier. Each dynode consists of a series of activated strips arranged like venetian blinds with alternating slopes between successive dynodes. As in the box and grid type, a fine mesh attached in front of each series of "venetian blinds" is maintained at the same potential as the dynode and prevents the electrons generated from falling back to the previous dynode. Because each dynode is maintained at successively higher potential or higher voltage, the electron flow is toward the collector.

Factors Affecting the Performance of Photomultiplier Tubes

Spectral Response

The photomultiplier tubes used for liquid scintillation counting must have a high sensitivity in the blue region of light because the liquid scintillators have blue emission. The spectral response of the photomultiplier is dependent on the composition of the photocathode, as shown in Fig. 4. These photomultiplier tubes using bialkali photocathodes (K_2CsSb) have been found to be particularly useful because of their excellent sensitivity in the blue region.

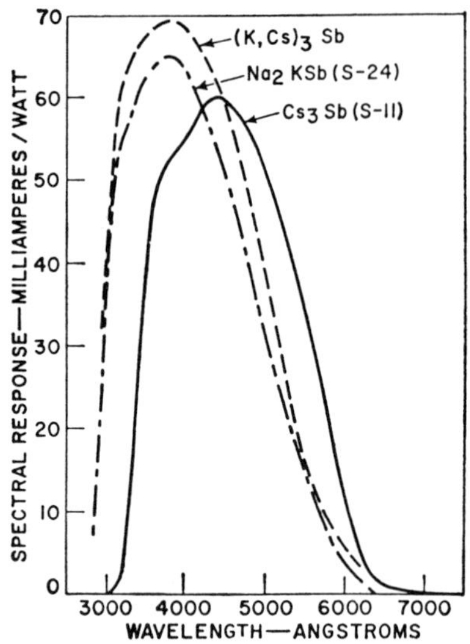

Fig. 4 Spectral response of photomultiplier tubes with various photocathode materials. (Courtesy of RCA.)

Magnetic Fields

The performance of all photomultiplier tubes may be influenced by the presence of a magnetic field because they generate photoelectrons that can be deflected by a magnetic field; similar to a charged particle in a mass spectrometer. In a liquid scintillation counter care must be taken to shield the photomultiplier tube from external magnetic fields to ensure constant and reproducible performance. The external magnetic field, however, can be manipulated electronically to modulate the performance of the photomultiplier tube (11). Under the application of normal external fields, no permanent damage results, although it is possible to cause slight magnetic polarization of some of the components of the tube resulting in the degradation of its performance. If magnetic polarization occurs, however, it is a simple matter to demagnetize the tube. For this reason, ferromagnetic material is avoided in tubes intended for use in liquid scintillation counters. An unusual application of this phenomenon will be discussed under "Efficiency Determination."

Temperature Effects

When measurements of gain are made on a photomultiplier tube with low current to avoid fatigue effects, it is found that they all have a small, but

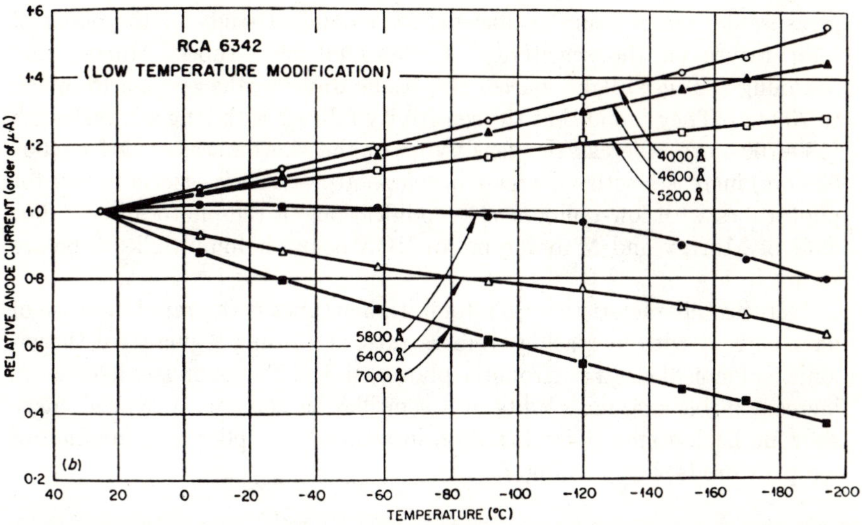

Fig. 5 Effect of temperature on the spectral response of a photomultiplier tube. (Courtesy of RCA.)

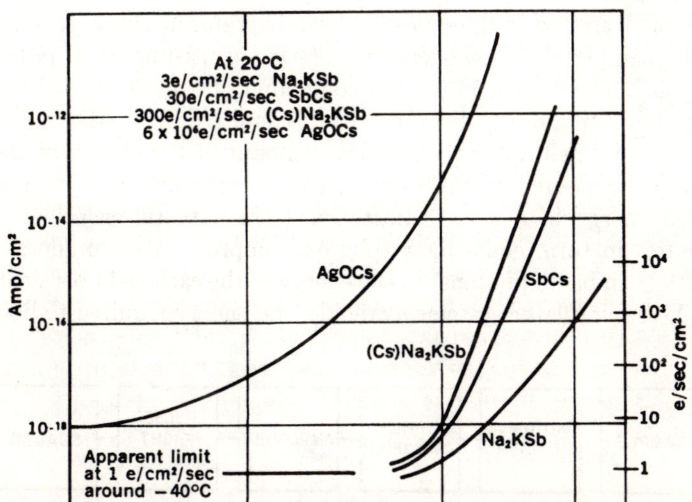

Fig. 6 Effect of temperature on the dark noise of photomultiplier tubes with various photocathode materials. (Courtesy of EMI.)

measurable, temperature coefficient. In a detailed study of the effect of temperature on the sensitivity of photomultiplier tubes, Murray and Manning (12) used light sources of various wavelengths ranging from 400 to 700 nm. They found that the sensitivity falls off with longer wavelength (580–700 nm), whereas the sensitivity to the shorter wavelengths (400–520 nm) increases with a decrease in temperature. This is especially true for phototubes with low-temperature modifications of the photocathode. The data of Murray and Manning on an RCA 6342 photomultiplier tube are shown in Fig. 5.

A critical characteristic of photomultiplier tubes is the rapid increase of dark current with increasing temperatures as a result of increased thermionic emission of electrons from the photocathode. This expresses itself as an increase in background counts in a scintillation counter. A typical curve showing background noise variation in a photomultiplier as a function of temperature is shown in Fig. 6.

THE SINGLE-TUBE LIQUID SCINTILLATION COUNTER

The measurement of radioactivity by means of a liquid scintillation counter is illustrated in the series of block diagrams in Fig. 7. It is assumed that the sample is a homogeneous solution consisting of a radioactive source, such as a carbon-14 labeled compound, dissolved in a solvent containing a scintillator. A beta particle is emitted from the radioactive compound. A small, but proportionate amount of the energy of the beta particle is transferred to the solvent, which transports and transfers the absorbed energy in a near-quantitative fashion to the scintillator. The scintillator, in turn, emits light (photons) approximately proportional in intensity (number of photons) to the energy of the carbon-14 beta particles. In the best scintillator systems available, the energy required to liberate a

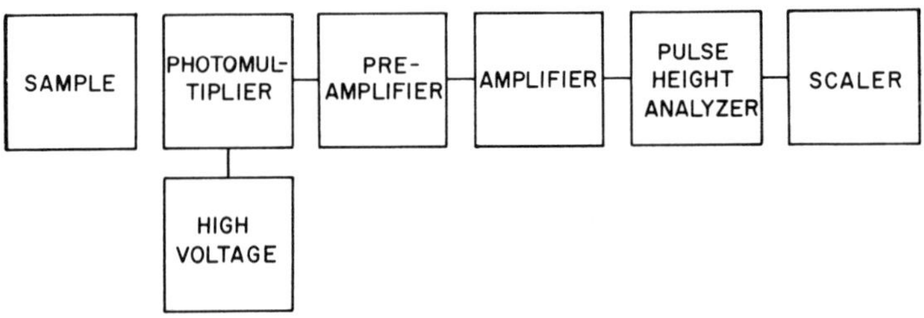

Fig. 7 Block diagram of a single-photomultiplier tube liquid scintillation counter.

single photon is about 150 eV, and an average carbon-14 beta particle, which has energy equivalent to 50 keV, produces about 350 photons. If we assume that all 350 photons impinge on the photocathode of the photomultiplier tube, about 25% of these result in primary photoelectron release by the photocathode. As explained above, these photoelectrons are now multiplied by the photomultiplier tube to produce a measurable electrical pulse. The resultant pulse is then amplified linearly by an amplifier to facilitate pulse height analysis. The pulse height analyzer is an electronic sorter that can be adjusted to accept electrical pulses within a selected range of pulse heights and reject all others. This is illustrated in Fig. 8. The bars, A through G, represent seven pulses of varying energies generated by the photomultiplier tube, the energy being directly proportional to the height of the bars, which are subsequently analyzed by a pulse height analyzer. The pulse height analyzer is set by means of two discriminators (lower and upper) that determine the size of the pulses to be passed on for further processing and reject all others. In our example, pulses A, C, and E are rejected because they are too small and pulses B, F, and G are rejected because they are too large. Only pulse D is passed on. This means that although seven pulses were generated at the photomultiplier tube, only one pulse was produced, in this case, by carbon-14, whereas the six rejected pulses were produced by background. Once the signal is passed on by the

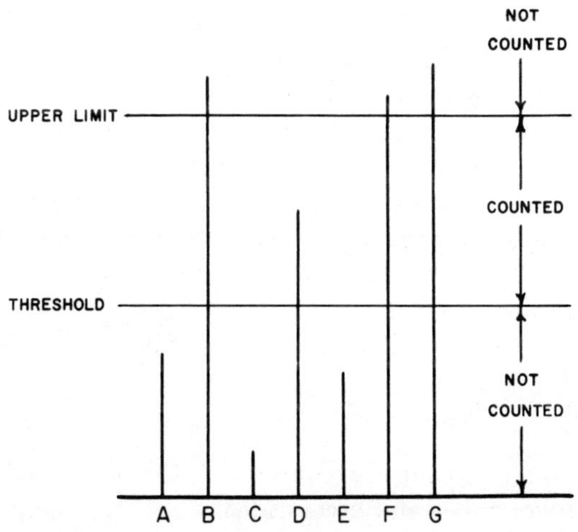

Fig. 8 Principle of pulse height analysis. Seven pulses of varying strength, A–G, are analyzed by a pulse height analyzer. Only those pulses which are greater than the threshold but less than the upper limit are accepted. Only one pulse, D, meets these criteria and is passed as an acceptable pulse (count). The bars represent arbitrary energy units: the higher the bar, the greater the energy.

pulse height analyzer as a legitimate carbon-14 pulse, the pulse is accepted by the scaler as a single count. The scaler is simply a tabulator of pulses.

Now, let us review the entire process. A beta particle is emitted from a carbon-14 source in solution. The energy of the beta particle is transferred proportionately by the solvent to the scintillator, which in turn emits some of this energy as light energy (photons). The emitted light causes the emission of photoelectrons from the photocathode of the photomultiplier tube and these photoelectrons are multiplied within the tube into a measurable electrical pulse. The electrical pulse is then amplified and passed on to the pulse height analyzer for analysis. The analyzer determines whether or not the impulse qualifies as a legitimate pulse. Because this is an acceptable pulse, it is passed on to the scaler, which records it as a count. This is the sequence of events that occurs in a liquid scintillation counter using a single photomultiplier tube.

THE TWO-TUBE LIQUID SCINTILLATION COUNTER

In principle, a liquid scintillation counter with a single photomultiplier tube should be adequate for laboratory use. In practice, however, the single-tube unit has a severe shortcoming that makes it of only limited use. The major restriction is determined by the noise characteristics of the photomultiplier tube because, in addition to emitting photoelectrons when exposed to light, the photocathode also emits electrons spontaneously (thermal electrons). The spontaneous release of these electrons is a function of temperature (see Fig. 5) and is markedly reduced by cooling. Even at reduced temperatures (e.g., 0°C), the spontaneous release of electrons can range from 10^3 to 10^7 pulses per minute at the tritium setting. Because these random pulses are equivalent to counts, this means that the background can range from 1000 to 10 million counts per minute for a single-tube unit set to count tritium.

Coincident Counting

The high background count is reduced to a tolerable level by the use of two photomultiplier tubes and a coincidence circuit. The photomultiplier tubes are specially selected for low-noise characteristics and matched for spectral response. The two photomultiplier tubes are so arranged that they look at the same sample and are connected to a coincidence circuit. The coincidence circuit is an electronic device that only passes signals arising simultaneously from both photomultiplier tubes. The coincidence circuit can reject, by design, random pulses that do not arrive within $3\text{–}8 \times 10^{-8}$ seconds of each other. The coincidence timing depends on the type of photomultiplier tube used and the judgment of the design engineer. Because the thermal electron emission from the photocathode is a random event, the accidental or coincident pulses resulting from thermal emission can be calculated from the following formula.

$$T = 2N_1N_2(R)$$

where T = coincident pulses per minute passed by the unit
N_1 = thermal noise rate in cpm in photomultiplier tube 1
N_2 = thermal noise rate in cpm in photomultiplier tube 2
R = resolution time of the circuit in minutes

For two photomultiplier tubes having thermal noise levels of 5000 cpm and 100,000 cpm and coincidence resolution times of 50 nseconds (5×10^{-8} seconds), the background contribution from thermal electrons is 0.83 cpm.

A diagramatic representation of a liquid scintillation counter employing a coincidence circuit is seen in Fig. 9. The second photomultiplier tube is used simply to monitor for coincidence. Therefore, any signal from the monitor tube is analyzed only by a lower level discriminator so that all signals above the minimum level are passed on to the coincidence unit. When the signal arrives at the coincidence unit from the monitor tube, an electronic gate is opened for a finite time, and any signal arriving from the analyzer tube and to the pulse height analyzer is allowed to pass on to the

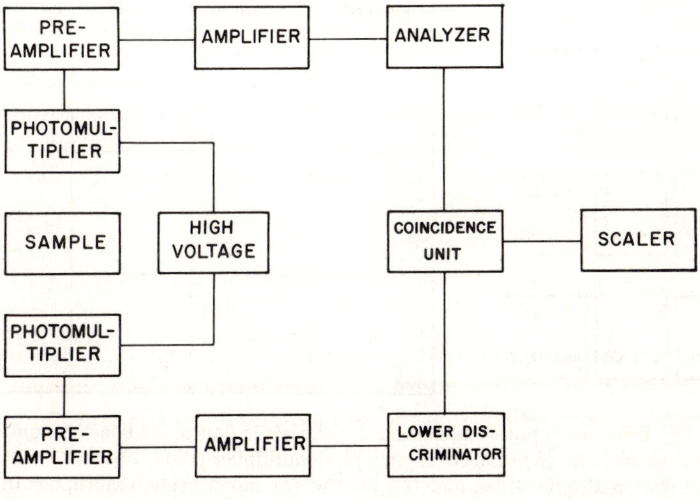

Fig. 9 Block diagram of a two-photomultiplier tube liquid scintillation counter with a single analysis channel.

counter during this time. Because pulses originating from the emission of a beta particle are seen simultaneously by both the monitor and analyzer photomultiplier tubes, these signals from the analyzer tube arrive simultaneously with those from the monitor tube at the coincidence gate and are counted. In contrast, the random background pulses originating in the analyzer tube do not pass the coincidence unit, even at very high background pulse rates.

Pulse Summation

In studying the logic of the design of a two-tube liquid scintillation counter (Fig. 9), it is apparent that the output from the second photomultiplier tube can be used for both analysis and for monitoring coincidence. This concept, known as "pulse summation," represents a real milestone in liquid scintillation spectrometer design and is now standard in the field. Electronically, it effectively doubles the signal input to the pulse height analyzer without any increase in the background noise level. The signal-to-noise ratio is, accordingly, improved 100% compared to the older design. From a practical point of view, pulse summation results in a more accurate pulse height analysis, as illustrated in Fig. 10.

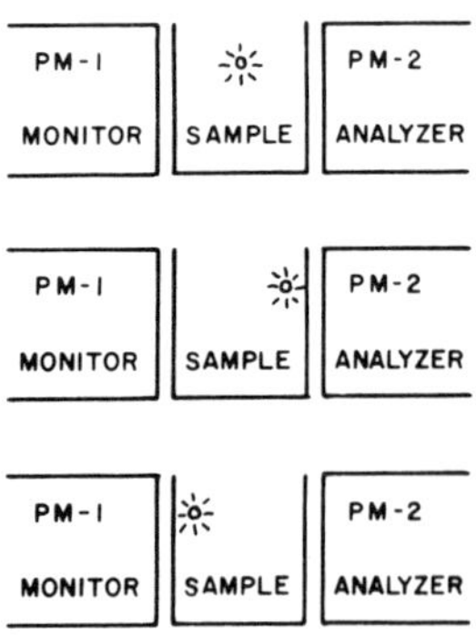

	Number of Photoelectrons Produced at Photocathode		Relative Pulse Height	
	Monitor PM Tube-1	Analyzer PM Tube-2	Unsummed PM Tube-2	Summed PM Tubes 1+2
A	5	5	5	10
B	1	9	9	10
C	9	1	1	10

Fig. 10 Principle of pulse summation. On the left is shown a radioactive sample that produces ten photons (A) in between two photomultiplier (PM) tubes, (B) next to the analyzer photomultiplier tube, and (C) next to the monitor photomultiplier tube. On the right is a table listing the hypothetical responses of the photomultiplier tubes with 100% photocathode efficiency.

Let us assume that a coincident liquid scintillation counter has two photomultiplier tubes, each with a photocathode efficiency of 100%, and is assaying a radioactive sample that is monoenergetic and results in the emission of ten photons per disintegration. We consider three hypothetical situations: (a) in which the ten photons are emitted between the two photomultiplier tubes such that five photons are collected at each photocathode, (b) in which the analyzer photomultiplier collects nine of the ten photons, and (c) in which the analyzer photomultiplier collects one of the ten photons. Without pulse summation, the number of primary photoelectrons resulting from a ten-photon emission varies from one to nine. This means that there is always a statistical distribution of pulse heights observed by a liquid scintillation counter without pulse summation even though the photon production by the radioactive source is constant. The pulse height depends on the position of the photon emission relative to the analyzer photomultiplier tube. In contrast, with pulse summation, the pulse heights are constant and, in this case, are always equivalent to that resulting from ten primary photoelectrons no matter where the photon emission occurs. It should be noted that the summing process must be arithmetic to achieve this constant output. If the summing process is applied to logarithmic pulses, the only advantage gained is a more favorable signal-to-noise ratio because the summing of logarithmic pulses is, in fact, a multiplication process and hence does not result in a constant product.

With pulse summation circuitry, it is important to match the photocathode efficiency of the two photomultiplier tubes to maintain linear summing response. In all high-performance liquid scintillation counters, the photomultiplier tubes are usually a selected matched pair. In counters in which equal voltages are applied to both tubes, the selection is made for near-equivalent photomultiplier gains at the same operating voltage. If the applied voltage to the photomultiplier tube can be varied individually, the photomultiplier gains can be matched by varying this voltage. Once the operating high voltage(s) is established, it is generally fixed for the life of the photomultiplier tubes. Even though the photomultiplier tubes are selected and matched, no two counters give identical performance in terms of reproducing an ideal spectral pattern for a given isotope. This is because of the difficulty in getting precise matching of photomultiplier tubes in terms of both spectral response (to various wavelengths of light) and photocathode efficiency. Pulse height adjustment, as described below and illustrated in Figs. 15 and 17, is accomplished by varying the amplifier gain linearly. In older instruments not employing pulse summation, pulse height adjustment is accomplished by varying the voltage impressed on the photomultiplier tubes and utilizing a fixed-gain amplifier. Experience has shown that this practice of varying gain on the photomultiplier tubes

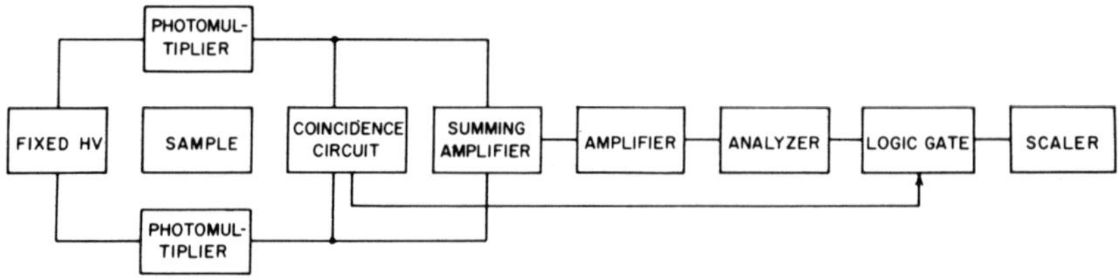

Fig. 11 Block diagram of a single-channel liquid scintillation counter with pulse summation. HV, high voltage supply for the photomultiplier tubes.

leads to nonreproducible performance because the photomultiplier tubes do not always return to their previous operating conditions. This procedure has therefore been abandoned in favor of using a fixed gain on the photomultiplier tube followed by either a variable amplifier or a fixed amplifier with an attenuator.

A block diagram of a liquid scintillation counter employing pulse summation is shown in Fig. 11. As described earlier, the high voltage on the photomultiplier tubes is fixed and the signals generated by the photomultiplier tubes are simultaneously monitored for coincidence and summed. When the coincidence circuit is triggered by the coincident pulses, the electronic gate is switched open to allow signals from the pulse height analyzer to pass for a finite time. While the gate is open, the same signal originating from the radioactive source is summed, amplified, and examined by the pulse height analyzer and passes through the open gate and is recorded by the scaler. A random noise pulse fails to trigger the coincidence circuit and is therefore unable to pass the electronic gate.

In Fig. 11, the coincidence gate is shown following the pulse height analyzer. Some commercial units are designed with the electronic gate (in this case, a linear gate) located between the amplifier and the pulse height analyzer. In either case, the result is the same. In summary, the modern liquid scintillation counter simply takes pulses from two photomultiplier tubes that are looking at the same sample and determines whether these pulses are coincident, sums them and checks the summed pulses for proper pulse height. The number of acceptable pulses is tabulated by the scaler.

Live Timing

A refinement known as "live timing" has been incorporated into the design of some liquid scintillation counters. In practice, live timing is a carryover from multichannel analyzer design of the vacuum tube era when the "fast" electronics capabilities of semiconductor devices and integrated circuits were unknown. When two pulses arrive at the pulse height analyzer at very close intervals, they can be "seen" as a single pulse resulting in a so-called "coincidence loss" of the second pulse. Furthermore, the two pulses that are "seen" as a single pulse can result in false pulse height analysis because the pulse heights may be additive. The live timer can eliminate this phenomenon by stopping the input to the pulse height analyzer as soon as it begins to process the first pulse and by reopening the input when the pulse is processed. Simultaneously, the clock is turned off during pulse processing and is turned on again after it is through. An analogous situation to live timing can be illustrated by considering a situation in which you are counting a steady stream of people entering a busy department store at the rate of 240 per minute. With a stop watch, you can count the people entering in 5-second intervals and record each count by hand. Obviously, while you are writing down the number of people counted, you cannot be counting the people who continue to enter the store. Therefore, during the recording period, or the count processing period, the clock is stopped. If the clock were not stopped during the count processing, the resulting count rate is inaccurate because some of the people entering the store during the processing period are missed. When you begin counting again, the clock is started simultaneously. Twelve such consecutive 5-second counting periods are equivalent to an accurate minute of continuous counting, but the total time (lapsed time) consumed in actually taking these 12 counts and recording the results is longer than 1 minute. This is the principle behind live timing. A modern liquid scintillation counter with a pulse processing time of 100 nseconds or less can easily accommodate a count rate of a million counts per minute with no perceptible coincidence loss (less than 0.2%). Sample count rates of this order of magnitude are seldom encountered in practice and, therefore, the incorporation of "live timing" seems superfluous. However, with the advent of the external standard method of quench determination (to be discussed below) in which high count rates are involved, "live timing" may be a desirable and, perhaps, a necessary feature. Live timing can also eliminate a phenomenon known as "after-pulsing," which occurs within the photomultiplier tube. An after-pulse can be generated by an initial pulse, and, if this after-pulse is allowed for analysis, it may result in a false count. By adjusting the processing period of the live timing cycle, sufficient time can be introduced to allow for any after-pulsing events to decay completely.

THE BETA SPECTRUM

An understanding of the nature of the beta spectrum of a typical weak beta emitter, such as carbon-14, and of the general principles of pulse

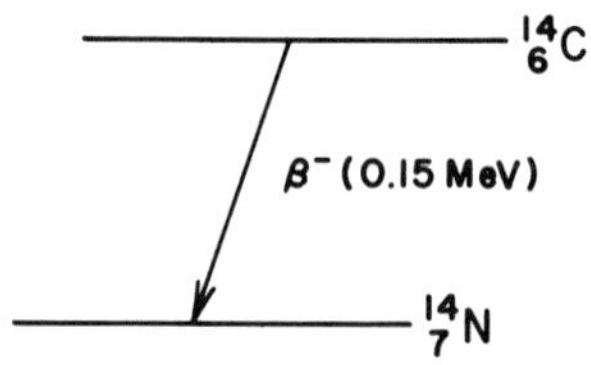

Fig. 12 Beta decay of carbon-14.

height analysis is required before we proceed with a discussion of counting techniques. Once they are well understood, these fundamentals can be applied to any of the other beta-emitting isotopes, such as tritium, sulfur-35, and phosphorus-32.

When carbon-14 disintegrates according to the scheme shown in Fig. 12 one expects the production of a beta particle, with a fixed energy of 0.156 MeV, and nitrogen. It has been found, however, that the disintegration of carbon-14 leads to the emission of beta particles with a range of energies varying from 0 to 0.156 MeV. Thus, a beta spectrum is said to be continuous.

Originally this nuclear transformation was not clearly understood and a second particle, the neutrino, was introduced to explain beta decay. It was postulated, for example, that all carbon-14 atoms that disintegrated gave up a beta particle and a neutrino particle that together contained 0.156 MeV of energy. Thus, the law of conservation of energy was satisfied.

The energy spectrum for carbon-14 can be visualized as shown in Fig. 13. The average energy of carbon-14 beta particles is roughly one-third of the maximum energy found for these beta particles, or about 0.05 MeV. As a

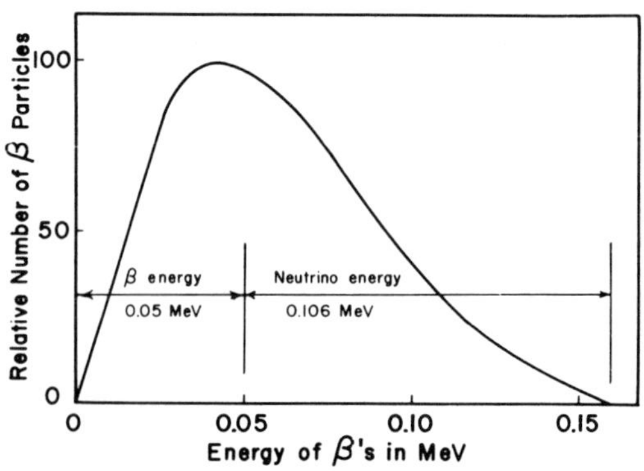

Fig. 13 Beta spectrum of carbon-14. The emission of an average beta particle (0.05 MeV) is accompanied by the emission of a neutrino with energy equivalent to 0.106 MeV. Together, they satisfy the requirement for the dissipation of energy equivalent to 0.156 MeV with every beta decay event.

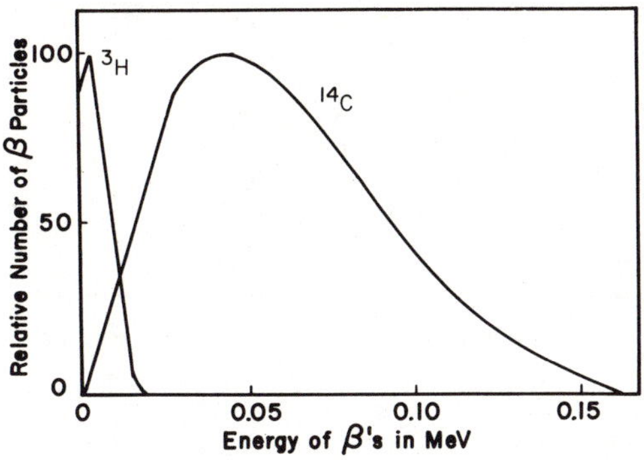

Fig. 14 Beta spectra of carbon-14 and tritium.

rule of thumb, the average energy of beta particles from any weak beta emitter is approximately one third of its maximum energy.

Tritium has a similar, but not exact, beta spectrum to that of carbon-14; the maximum energy of a tritium beta particle is 0.018 MeV. The tritium spectrum superimposed on the carbon-14 spectrum is shown in Fig. 14.

PULSE HEIGHT ANALYSIS

We are now in a position to discuss pulse height analysis, keeping in mind the nature of the beta spectrum. The pulse height analyzer has been described above as an electronic sorter. Alternatively, it can be regarded as an electronic window that can be opened or closed to "see" electronic pulses that fall within the window opening. The width of the window opening can be expressed arbitrarily as relative volts having a range of zero to 100. In a single-channel liquid scintillation counter, assume that we set the lower discriminator of the pulse height analyzer at 10 V and the upper discriminator at 60 V and place a carbon-14 sample into the counter. If the amplifier gain is slowly and steadily increased (or the attenuation decreased) the count rate first gradually increases, reaches a maximum, and then starts to decline. This phenomenon is illustrated in Fig. 15. At gain position 1 none of the carbon-14 beta particles have sufficient amplification to be seen by the pulse height analyzer. At gain 4, the pulses are now amplified and only the beta particles with the highest energy are seen and passed by the pulse height analyzer. At gain 6, all the beta particles with near-average energy content are accepted, whereas those with very low or very high energies are excluded. At gain 60, only the low-energy end of the carbon-14 beta spectrum is in the window. The efficiency of the counter is

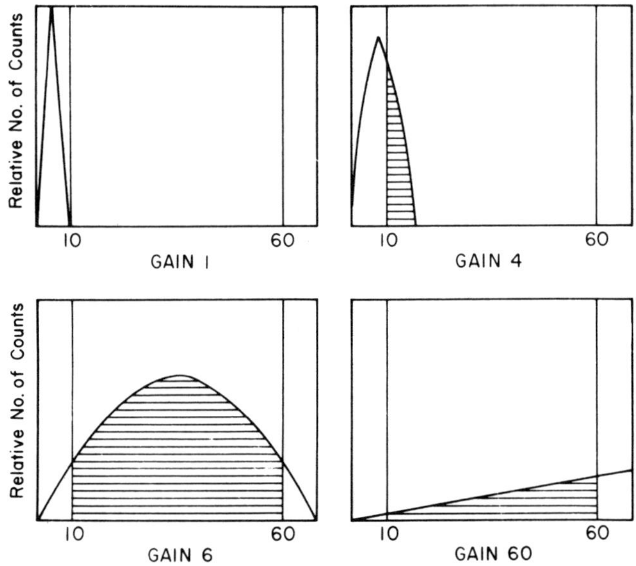

Fig. 15 Carbon-14 spectra in a 10–60 pulse height analyzer window (100 maximum) at various amplifier gains.

the percent of the beta spectrum seen by the analyzer, which, in turn, is determined by the overall gain of the counter and by the discriminator settings of the pulse height analyzer, assuming that the sample is ideal (unquenched). The approximate efficiency for carbon-14 for the overall gains described above may be as follows: at gain 1, 0%; at gain 4, 40%; at gain 6, 70%; at gain 60, 10%.

BALANCE POINT COUNTING

The balance point is an arbitrarily determined amplifier gain position at which the beta spectrum under consideration is nearly symmetrical in the analyzer window and is at the gain setting that gives the maximum efficiency. This is illustrated by an idealized carbon-14 beta spectrum at gain 6 (Fig. 15). The advantage of this position is that it can accommodate any small shift in the curve caused either by slight fluctuations in line voltage or by the performance of electronic components. A small decrease in gain results in a shift of the spectrum to the left, as shown by the dashed curve in Fig. 16. It can be seen that there is a loss of counts on the right side and a gain of counts on the left. The net change is small. If the curve shifts to the

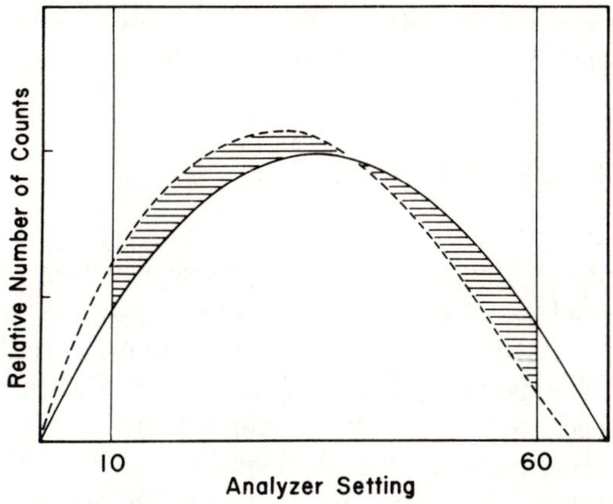

Fig. 16 Effect of amplifier gain change on the carbon-14 spectrum at the balance point. The solid line depicts an idealized carbon-14 spectrum at balance point. The dashed line shows the spectrum when there is a small change (loss) in amplifier gain. The vertical bars represent the position of the analyzer settings. The hatched areas represent the change in loss in the number of pulses lost and gained after a small amplifier gain. This is explained in the text.

right, the same considerations apply. Balance point operation for each type of sample, therefore, represents the most stable counting condition. Although balance point counting is essential when the old vacuum tube units are used, the stability of modern day linear amplifiers and electronics has allowed reproducible counting off balance point.

The tritium spectrum, superimposed on the carbon-14 spectrum, is shown in Fig. 17. Tritium pulses are approximately ten times weaker than those of carbon-14 and in order to get tritium pulses into the window more amplification is required. The counting considerations discussed for carbon-

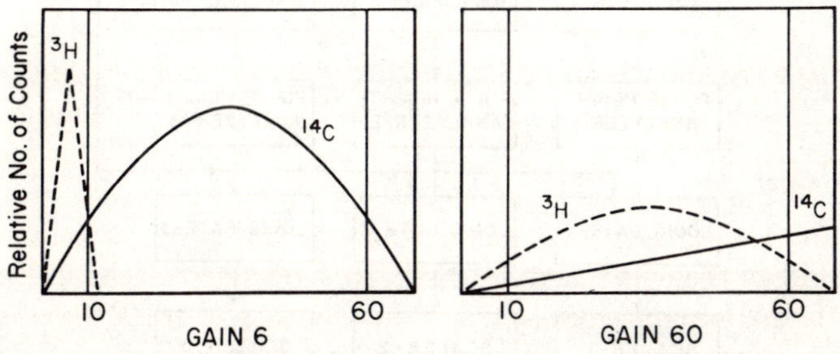

Fig. 17 Carbon-14 and tritium spectra in a 10–60 pulse height analyzer window (100 maximum) at various amplifier gains.

14 can also be applied to tritium. When the tritium pulses are seen at balance point (gain 60), between 55 and 65% efficiency is obtained.

DOUBLE-ISOTOPE ANALYSIS

One of the earliest applications of a liquid scintillation counter was for double-isotope analysis of a steroid. In the study by Avivi *et al.* (13), a homemade liquid scintillation counter was used to measure the total radioactive content of experimental steroid samples containing both tritium and carbon-14 because their counter could not distinguish between these two isotopes. A Geiger counter was then used to exclude tritium! The early commercial counters, however, could distinguish between tritium and carbon-14, and this facility stimulated the development of double-isotope analysis. However, the low counting efficiencies for tritium (about 10–15%), poor isotope discrimination, and poor instrument stability made this

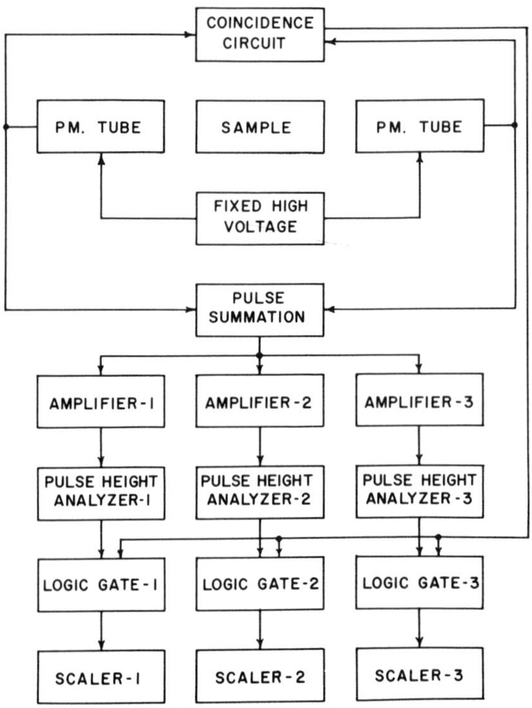

Fig. 18 Block diagram of a three-channel liquid scintillation counter with variable linear amplifiers and pulse summation. PM., photomultiplier.

analytical technique a very delicate procedure. Balance point counting was a necessary requirement with these early instruments, but with modern counters the accurate counting of doubly labeled samples does not present any serious difficulties.

One of the most useful features of a liquid scintillation counter is this ability to differentiate between isotopes that emit beta particles of varying energies and to count them simultaneously at optimum counting conditions for each isotope. To facilitate double-isotope analysis, most liquid scintillation counters today are multichannel units containing two or three independent sample analysis channels. A block diagram of a three-channel liquid scintillation counter with pulse summation and linear amplifiers is shown in Fig. 18. With such a unit, tritium can be analyzed in channel 2 and carbon-14 in channel 1. The third channel can be used for a variety of purposes, such as for efficiency determination by the sample channels ratio method.

In the following discussion of double-isotope analysis, a hypothetical sample containing both tritium and carbon-14 is assumed and the liquid scintillation counter characteristics are those described by Figs. 15 and 17. Only two channels of the counter are used for analysis in the following method, which is commonly used for this purpose.

Channel 1 is set at the carbon balance point (gain 6, 10 to 60-V window) and channel 2 is set at the tritium balance point (gain 60, 10 to 60-V window). A carbon-14 standard and a tritium standard, dissolved in the same solvent and containing the same scintillator used for experimental samples, and a reagent blank are counted first. Then, the experimental samples are counted. The contribution of each isotope to the total counts observed in test samples is calculated with the use of the following simultaneous equations.

C = carbon dpm in sample (dpm = disintegrations per minute)
H = tritium dpm in sample
c_1 = carbon-14 efficiency in channel 1 (i.e., 100% = 1.00)
c_2 = carbon-14 efficiency in channel 2
h_1 = tritium efficiency in channel 1
h_2 = tritium efficiency in channel 2
N_1 = net total observed counts in channel 1
N_2 = net total observed counts in channel 2

Then

$$N_1 = C(c_1) + H(h_1) \tag{1}$$

and

$$N_2 = C(c_2) + H(h_2) \tag{2}$$

If the above equations are solved for C and H, it is found that

$$C = \frac{N_1 - N_2(h_1/h_2)}{c_1 - c_2(h_1/h_2)} \tag{3}$$

$$H = \frac{N_2 - N_1(c_2/c_1)}{h_2 - h_1(c_2/c_1)} \tag{4}$$

It can be seen in Fig. 17 that at gain 6 very little of the tritium spectrum is in the carbon window. By simply raising the lower window of the pulse height analyzer a few relative volts, all the tritium can be excluded from the carbon-14 window (channel 1). Under these conditions, the equations are simplified as follows.

$$N_1 = C(c_1) \tag{5}$$

and

$$N_2 = C(c_2) + H(h_2) \tag{6}$$

If these equations are solved for C and H, it is found that

$$C = N_1/c_1 \tag{7}$$

$$H = [N_2 - C(c_2)]/h_2 \tag{8}$$

Equations (3) and (4) are basic equations that are applicable to any liquid scintillation counter being used for double-isotope analysis. If a single-channel counter is used, c_1 and h_1 are the efficiencies of the carbon-14 and tritium standards at the carbon-14 balance point and c_2 and h_2 are the efficiencies at the tritium balance point.

The isotope exclusion method of double-isotope analysis illustrates a fundamental point, which is that the lower isotope can always be excluded from the higher energy isotope but the higher energy isotope can never be excluded from the analysis of the lower energy isotope. This follows from the nature of the beta spectrum, which is distributed from almost zero energy to a finite value. An advantage of pulse summation is that in double-isotope analysis approximately 70% of the carbon-14 spectrum lies above the tritium spectrum in an unquenched sample. Without pulse summation only 40% lies above the tritium spectrum.

In older, single-channel instruments with three discriminators or a divided window, it is possible to count tritium and carbon-14 in a single operation. This is not the most accurate way to count such samples but it may be advantageous as a method to conserve counting time. In these instruments, the dynamic range of the pulse height analyzer is limited in order that when the carbon-14 spectrum is in the window very little of the tritium spectrum can be analyzed. In order to get enough tritium into the

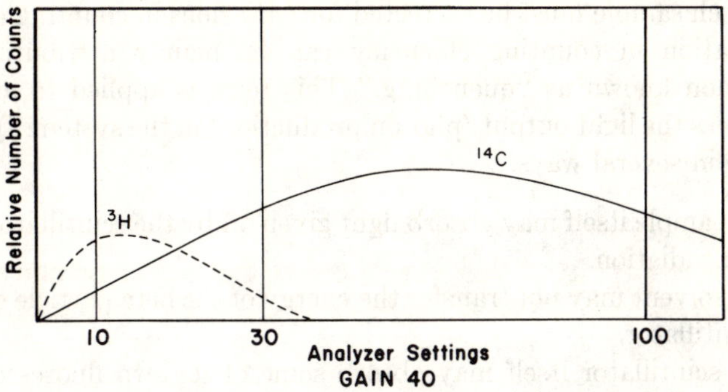

Fig. 19 Carbon-14 and tritium spectra in a single-channel liquid scintillation counter with three discriminators set for double-isotope analysis.

window, the gain must be increased to a point where most of the carbon-14 spectrum is beyond the upper window of the pulse height analyzer and neither spectrum is at balance point. As most of these counters are vacuum tube units, counting off balance point results in a high degree of variation (5–10%) between replicate counts. A typical counting setup for this method is shown in Fig. 19. An intermediate gain between carbon-14 and tritium balance points is selected (i.e., gain 40). One window is set to count tritium efficiently and carbon-14 inefficiently (10 to 30 V window), and the second window is set to count carbon-14 efficiently and tritium inefficiently (30 to 100 V window). To calculate the carbon-14 and tritium contributions to the sample counts, Eqs. (3) and (4) are used. As before, c_1, h_1, and N_1 are efficiencies and net counts in one window, whereas c_2, h_2, and N_2 are efficiencies and net counts in the second window.

QUENCHING

In the discussions on pulse height analysis it has been pointed out that most of the currently available commercial counters are instruments allowing reproducible counting when used off the balance point for the isotope. It is rare, however, that a series of diverse experimental samples can be prepared with the same degree of uniformity. Most samples therefore exhibit variations in counting efficiency. These variations may be caused by the nonuniformity of the vials, the amount of oxygen dissolved in the solvent, or the variable absorption of photons by the sample itself. Provided an effort is made to prepare samples uniformly, the variation in sample counting efficiency among a group of samples is usually small enough (within 0.5%) to be ignored; however, if precise determinations are required, each sample must be corrected for variations in counting efficiency. The variation in counting efficiency can be mainly attributed to the phenomenon known as "quenching." This term is applied to any factor that reduces the light output (photon production) in the system. Quenching can occur in several ways.

1. The sample itself may absorb light given off by the scintillator or some of its own radiation.
2. The solvent may not transfer the energy of the beta particle efficiently to the scintillator.
3. The scintillator itself may absorb some of its own fluorescence.
4. Chemical interaction of the components contained in the counting solution may result in reduced photon output.

It is the phenomenon of quenching that makes the prediction of performance for any given compound uncertain and that is responsible for the empirical nature of the sample counting procedures available. It is by far the most serious problem encountered in liquid scintillation counting.

The energy of the beta particle emitted by the sample is absorbed by the scintillator and the number of photons released is directly proportional to the energy of the beta particle absorbed. The number of photons striking the photocathode of the photomultiplier tube varies, therefore, with the isotope responsible for the production of these photons. The brighter the "light" flash of the scintillator, the greater the number of photons emitted. If 400 photons are produced by a carbon-14 beta particle of near-average energy, about 200 photons should strike each photomultiplier tube in a coincidence counter. Because the quantum efficiency of the photomultiplier tube is about 25%, the 200 photons striking the photocathode produce about 50 electrons for subsequent multiplication by the dynode chain. Quenching within the sample, by any of the mechanisms mentioned above, reduces the number of photons reaching the photocathode. This loss of photons results in a reduction of pulse height. The original pulse height can be restored by increasing the overall amplification of the system. Whenever quenching occurs, it is still theoretically possible to count each disintegration as long as a minimum of four photons strikes each photomultiplier tube to generate a photoelectron. If quenching is so severe that less than eight photons are emitted by the sample, the disintegration is lost to the counter. An average tritium beta particle produces about 40 photons in a toluene scintillator solution. If the above considerations are now applied to tritium, an unquenched average tritium disintegration results in the production of about five photoelectrons in each photomultiplier tube. At such a low photoelectron production, the discrimination of tritium against the thermal noise of the photomultiplier tube is difficult. This explains the lower counting efficiency for tritium when compared to that of carbon-14 and also explains why only moderate quenching in tritiated samples is needed to result in an irretrievable loss of counts.

A simple test for quenching by a radioactive compound is to count a given amount of sample and then to double the amount of the sample. If quenching is absent, a doubling of the counts is observed. If quenching caused by the sample is present, the counts are less than double that of the original sample. It is assumed that the radioactive sample being tested is uniform and that the solvent and scintillator concentrations remain constant.

EFFICIENCY DETERMINATION

The determination of quenching is synonymous with the determination of sample counting efficiency in the field of liquid scintillation counting. It must be kept in mind that the counting efficiency of the counter itself does not change once the amplifier gains and pulse height analyzers are set. Quenching occurs only in the sample, and the maximum sample counting efficiency for a particular electronic setting of the counter is that obtained by an unquenched sample. The most common methods used to determine sample counting efficiency are (a) the internal standard method, (b) the channels ratio method, and (c) the external standard method.

The determination of counting efficiency allows the conversion of counting data into absolute units (by dividing the net counts per unit time by the efficiency). The common unit is disintegrations per minute, usually abbreviated dpm. Counting data expressed in this unit eliminates any ambiguity, such as that associated with counting activity expressed as counts per minute. It is recommended that all counting data be universally expressed in absolute units.

Internal Standard Method

The classical method for determining counting efficiency is the internal standard method. In this procedure, the sample is first counted, a calibrated amount of a nonquenching radioactive standard is added, and the sample is recounted. The additional counts produced by the internal standard are used to compute the counting efficiency. The most commonly used internal standard is labeled toluene, although such proved nonquenchers as cholesterol or *n*-hexadecane are also satisfactory. Benzoic acid has been used as an internal standard but because it is a quenching agent its continued use is not recommended. The internal standard is added in a small volume (0.1 ml or less) so as not to alter the counting characteristics of the original sample. The disadvantages of the internal standard method are as follows.

1. A pipetting error is introduced when the internal standard is added.
2. Additional counting time is required.
3. Opening the vial to add the internal standard and then recapping it may produce a sample that is different from the original because of water absorption, sample evaporation, and variations in oxygen quenching.
4. Rotation of the vial as it moves round the conveyor belt means that the geometry of the vial in relation to the photomultiplier tubes is probably different after the internal standard has been added.
5. Contamination of the sample by the internal standard may occur.

In spite of these shortcomings, the internal standard method remains the least controversial method of efficiency determination.

Channels Ratio Method

This method is based on the fact that the pulse height spectrum is always displaced when quenching occurs. In a two-channel instrument assaying carbon-14, it is theoretically possible to set one channel to include all beta particles having energies from 0 to 156 keV, which would be equivalent to a counting efficiency of 100% (A in Fig. 20). The second channel, B, can be set to count all beta particles having energies from 50 to 156 keV, which would be equivalent to a counting efficiency of 50%. If the channels ratio is arbitrarily defined as the narrow-window efficiency divided by the wide-

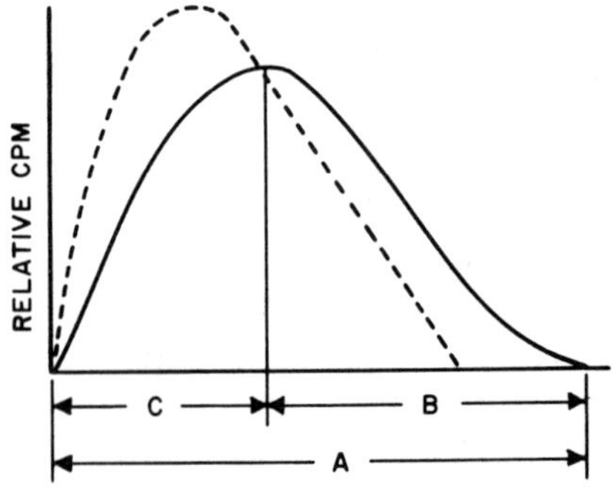

Fig. 20 Carbon-14 spectra. The solid curve represents unquenched carbon-14. The dotted curve represents quenched carbon-14. A, B, and C represent various window widths of pulse height analyzers that can be used for the channels ratio method of efficiency determination.

window efficiency, the ratio is always a number less than one. With a two-channel instrument set in the above manner, nonquenching samples all containing various amounts of carbon-14 material always have a channels ratio of 0.5. The ratio indicates the counting efficiency in the narrow window. The absolute activity (dpm) of each sample is then obtained by dividing the net counts per minute appearing in the narrow window by the counting efficiency determined from the ratio. The accuracy of the channels ratio method is a function of the accuracy of the standards used to prepare the ratio plot and of the stability of the instrument. A typical channels ratio curve for the conditions outlined is shown in Fig. 21. In this example, the ratio decreases as quenching increases. It must be pointed out that any ratio can be used for the channels ratio; that is, A/B, B/C, or C/B. A standardized series of quenched samples containing either carbon-14 or tritium for establishing a channels ratio versus efficiency curve is commercially available. However, these commercial quenched standards are only useful if the counting vial, cap, solvent, and scintillator(s) are identical to the experimental samples. Valid quench curves can only be constructed by using the same solvent(s), scintillator(s), vials, and caps as those to be used in the experiment. The major weakness of the channels ratio technique is in the analysis of samples containing small amounts of radioactivity, for which the time required to accumulate a statistically acceptable number of counts may become prohibitive.

The channels ratio method is independent of the volume of the sample and has the following advantages over internal standardization.

1. No additional time is required for efficiency determination as is required in the internal standard method.

2. The sample does not require rehandling, thus eliminating accidental loss of the sample because of carelessness.

3. The sample is not contaminated by the addition of a radioactive

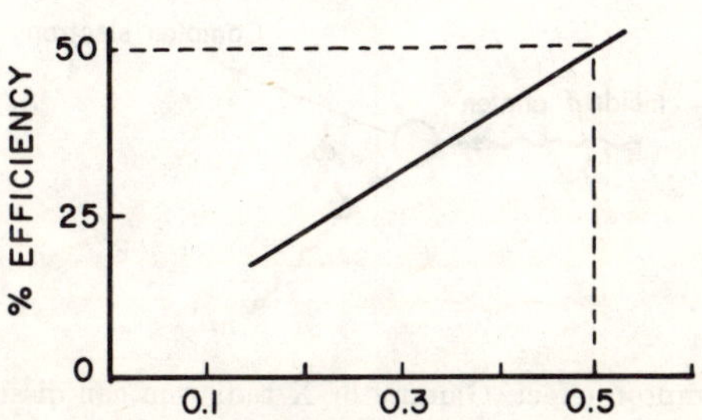

Fig. 21 A typical channels ratio curve for efficiency determination. A and B are pulse height analyzer windows shown in Fig. 20.

internal standard. Therefore, should the need arise, the sample may be recovered for further analysis.

4. A small change in volume that may change the degree of quenching results from the addition of the internal standard. Furthermore, precautions must be taken to thermally equilibrate the sample before adding the internal standard to prevent moisture from condensing in the vial. These possible sources of error are avoided in the sample channels ratio method.

External Standard Method

The external standard method for efficiency determination was introduced commercially in 1964. This method utilized the fact that irradiation of both the sample vial and solution by a gamma source produced Compton electrons. (Fig. 22). The basic premise was that these Compton electrons behaved as the beta particles emitted by the radioactive sample in solution and any quenching in the sample resulted in a proportionate quenching of the Compton electrons.

The first application of the external standard was carried out in the following manner. A radium-226 or a cesium-137 pellet was used as the gamma source. After the sample was counted, the gamma source was automatically positioned near the sample vial and the sample counted for an additional minute. The counts from the Compton electrons generated by the irradiation of the sample by the external standard were monitored by an additional pulse height analyzer set to "see" only the very high energy particles. These counts, exclusively from the external standard, were then directly related to the counting efficiency. The external standard was highly active so that a statistically significant count was obtained with an

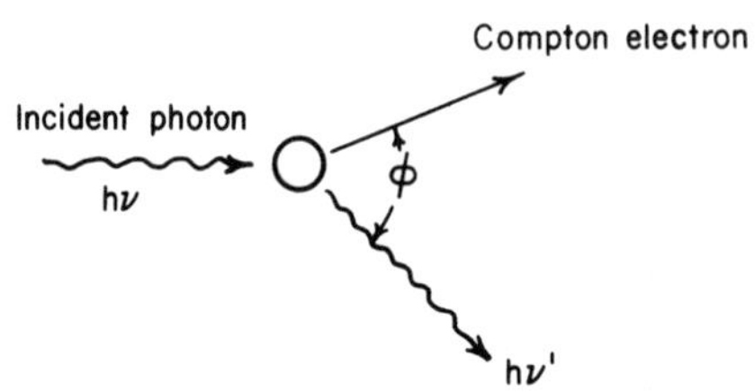

Fig. 22 The Compton effect. Gamma or X radiation can dissipate the energy of a photon via the Compton effect. The incident gamma or X radiation photon with an energy $h\nu$ interacts with a planetary electron, causing the ejection of a recoil electron called the "Compton electron." This process absorbs some of the energy of the incident photon. The incident photon is simultaneously deflected at an angle, ϕ, and emerges as a scattered photon of lesser energy, $h\nu'$.

additional minute of counting. This novel approach for efficiency determination had all the advantages mentioned for the channels ratio method over the internal standard method with the exception of volume dependence. Its three greatest assets over the samples channels ratio method were as follows.

1. Counting efficiency can be determined in any sample regardless of its radioactive content.
2. It can be used to determine efficiencies in double-isotope analysis regardless of the isotope ratio, i.e., carbon-14 to tritium.
3. The determination can be made quickly.

The external standard method, however, is not without its shortcomings.

1. The basic process that produces Compton electrons is the interaction of gamma-ray photons with the planetary electrons of the atoms and is mass dependent. The external standard method therefore, as used in liquid scintillation counting, is dependent on the volume of the sample. This volume dependency can be minimized by doing a ratio plot of the external standard spectrum in manner similar to that described above for the channels ratio method. A novel method for volume correction utilizes both a channels ratio determination and a second gamma emitter to monitor volume.
2. The reproducibility of the method depends on the maintenance of constant geometry between the external gamma-ray source and the sample vial. This is difficult in an automatic system, for after being cycled by the sample belt the sample is seldom returned to exactly the same position on the sample elevator for recounting.
3. The Compton electrons produced by the external standard are energetic electrons and do not mimic exactly the behavior of weak beta particles in the sample itself.
4. The external standard cannot always be used to determine efficiency in a heterogeneous counting mixture.

LESSER PULSE HEIGHT ANALYSIS

Although Bush (14) has reported that the sample channels ratio curve for efficiency determination can correct for either color or chemical quenching, experience has shown that this is not so for quench correction curves derived from external standard ratios. For equivalent degree of quench, the pulse height spectra are generally sharper and less diffuse for chemical quench than for color quench (Fig. 23A). This is reasonable because chemical

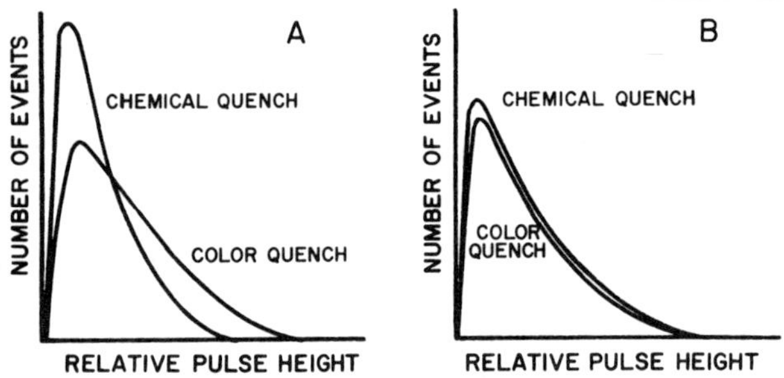

Fig. 23 Typical spectra of two carbon-14 samples quenched to equivalent counting efficiencies by chemical quenching and by color quenching. (A) The spectra plotted using pulse summation and linear amplification and (B) the spectra plotted using the lesser pulse height and linear amplification.

quench occurs before the generation of photons by the scintillator, and its severity is simply a function of quencher concentration. Color quench, in contrast, occurs after photon production and depends on the path length of the photon through the colored solution. For example, when a single beta-decay event occurs in a colored solution, a number of photons of equivalent energy are produced initially. If the decay event occurs anywhere but the center of the vial, the photons that give rise to a coincident event produce unequal pulse heights in the two photomultiplier tubes. Those photons which travel the greater distance are less energetic, having experienced the color quenching environment longer, and give rise to the smaller pulse height. The lesser pulse height is always generated by the phototube farthest from the beta decay event. The diffuse nature of the color quench curve (Fig. 23A) may then be a reflection of the inability of the pulse summation system to cope with the fact that pulses being summed have unequal color quench components. Laney (15) has reasoned that the analysis of only the smaller of the two pulses that produce coincident events can give a more accurate estimation of color quench and lessen the disparity between color and chemical quench correction curves using external standard ratios. This technique is called "lesser pulse height analysis." Conceptually, the method returns to the original single-analyzer tube approach, with the important distinction that either single photomultiplier tube can provide the lesser pulse height, thus reducing the spread of values by a factor of two over the original system. (In the example given in Fig. 10, the spread in values for the lesser pulse height analysis is one to five instead of one to nine for the old single-analyzer system.) However, the advantage of pulse summation is sacrificed and it appears that there should be some deterioration in spectral separation, although this is disclaimed by

Laney (15). It should be stressed that lesser pulse height analysis is an attempt to utilize the external standard ratio to generate a near-universal quench correction curve. This approach is unproved at this writing. It should be mentioned that a conventional system using pulse summation and an external standard ratio generated by radium-226 and americium-241 can satisfactorily monitor chemical and color quench with a single curve over a wide range of quench. Divergence is encountered with extreme color quenching but this is true for all quench correction systems, including lesser pulse height analysis.

FIGURE OF MERIT

In the field of liquid scintillation counting, the performance of a given counter is often expressed as a figure of merit. This procedure for expressing instrument performance has arisen from the perennial question of how long a sample should be counted to be statistically acceptable. The limits of acceptability may differ from experiment to experiment but the basis for this statistical determination is fixed. The time required to achieve this acceptable statistical number is a function of the efficiency and of the background of the counter. The figure of merit is (E^2/B), where E is the efficiency expressed as a percent and B is the background in counts per minute at this efficiency. The higher the figure of merit, the less counting time is required to achieve the statistical accuracy desired. In new counters, a figure of merit in excess of 350 for carbon-14 and over 150 for tritium for a standard toluene sample prepared in air may be expected. The figures of merit quoted by manufacturers for tritium range from 120 to 200 and for carbon-14 from 300 to over 400. These figures, however, are obtained using unquenched, oxygen-free standards. In practice, the figures of merit obtained with most laboratory samples are much lower.

LOGARITHMIC LOGIC COUNTERS

The first commercial liquid scintillation counter utilizing logarithmic amplifiers was introduced in 1964. A block diagram of a logarithmic unit is shown in Fig. 24. An advantage claimed for this design was simplicity because only a single amplifier was used for all analyzer channels. It was also claimed that the elimination of the variable attenuators required in the

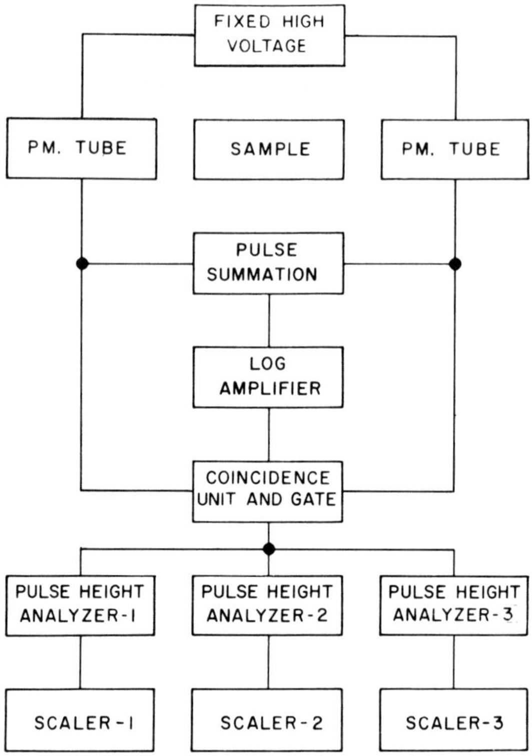

Fig. 24 Block diagram of a three-channel liquid scintillation counter using logarithmic amplification. PM., photomultiplier.

linear amplifier design (Fig. 18) resulted in more stable and reproducible performance.

The spectra of tritium, carbon-14, and chlorine-36 plotted from a logarithmic unit appear as shown in Fig. 25. The abscissa represents the 100 relative volt window of the pulse height analyzer. When phosphorus-32 is plotted on this graph, it will peak at about 80 relative volts. It can be seen that with the log unit the isotopes of interest in biochemistry, tritium, carbon-14, sulfur-35, and phosphorus-32, can be discriminated by merely adjusting the pulse height analyzers, that is, by merely adjusting the upper and lower windows of a given channel. Conceptually, the log amplifier approach to liquid scintillation counter design produces the simplest machine to use from the operator's point of view. The only adjustment required for analyzing any isotope is that of the pulse height analyzer windows. The acceptance of log units, however, has been slow. The overall

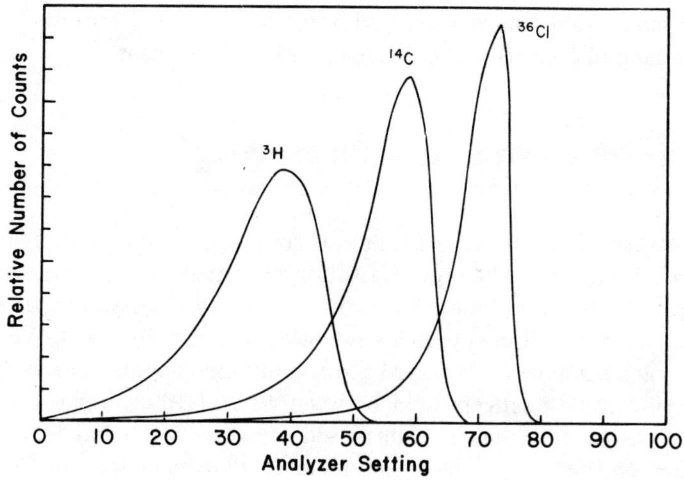

Fig. 25 Spectra of tritium, carbon-14, and chlorine-36 determined in a liquid scintillation counter with logarithmic amplification.

stability and reproducibility in performance has not been fully established as evidenced, for example, by the number of publications of double-label work in the steroid field, in which counter stability is mandatory because of the low count rates usually encountered. It is difficult to maintain the kind of stability required in scintillation counting with logarithmic amplifiers over a three- to four-decade relative energy range. Furthermore, inspection of the spectra in Fig. 25 shows that the carbon-14 spectrum above tritium occupies less than 20% of the maximum pulse height analyzer window width. This means that the discriminator settings must be very accurate, easily reproducible, and stable to be useful. In a linear system, the carbon-14 spectrum lying above the tritium can be made to occupy about 60–70% of the maximum pulse height window opening, and this allows more latitude in reproducing the settings and maintaining pulse height analyzer stability. This is of particular importance in double-isotope analyses.

A logarithmic spectrum can also be obtained by operating the photomultiplier tubes in such a way as to give a logarithmic rather than a linear response. The logarithmic output of the photomultiplier tubes can then be summed, amplified by a linear amplifier, monitored for coincidence, and finally analyzed by the pulse height analyzers. This approach is identical to that shown in Fig. 24, except that the amplifier is linear and that the summing circuit adds the logarithmic signals. Note, however, that in a linear system summing involves addition of the signal outputs, whereas the summing of logarithms is equivalent to multiplication.

AUTOMATION AND DATA PROCESSING

The utility of the liquid scintillation counting technique in generating samples has catalyzed the rapid transition in instrumentation from manual counters to automated units. Automation consists of an automatic sample changing device combined with a readout device, usually a simple lister, to identify each sample and to record the accumulated counts in each channel. The early sample changers hold between 40 and 100 samples but, more recently, the capacity has been increased to between 100 and 450 samples, depending on the model. These samples are held in some form of an endless belt or an array of removable sample trays. When samples are analyzed automatically, two limiting parameters are set: the maximum counting time and the maximum number of counts to be accumulated. Limiting the counting time assures a maximum counting time for low-activity samples and limiting the accumulated counts allows shorter counting times for high-activity samples while preserving the desired statistical accuracy. As an added convenience, each model employs its own method of selecting groups of samples to be analyzed while omitting all others.

The capability for preselecting time, counts, and groups allowed efficient use of available counting time. The first attempt at data processing took the form of a direct substitution of a mechanical calculator for the lister. The calculator divided the accumulated counts by the elapsed time and printed counts per minute per channel. The next step was to automatically correct for background and this took form in a calibrated oscillator set to pulse at the background rate. The principle behind this innovation was that periodic pulses from the oscillator would null or cancel coincident sample pulses. At high count rates this concept worked, but at low count rates it proved unsatisfactory because radioactive decay is random and the pulses from the oscillator are periodic. The oscillator therefore failed to cancel most sample pulses, leading to a large cumulative error. Although improvements in the device were made, this method of correcting for background was eventually abandoned. An alternative method of background correction involved predetermining the background rate. The expected cumulative background counts for a given constant counting time were then entered into the appropriate background register. For example, if the background was 20.3 cpm and the sample counting time 10 minutes, 203 counts were entered into the background register of the proper channel. Then, for each sample counting started at 999,797 counts so that the first 203 counts would set the scaler to zero because the capacity of the scaler was six digits, or 999,999. The millionth count caused the scaler to read six zeros. This method of subtracting background automatically was obviously only valid for samples that counted the full 10 minutes.

Today, the more sophisticated counters have incorporated not only the capability of correcting for background electronically, irrespective of counting time, but also the ability to compute and print the net counts per minute per channel as well as the desired channels ratio for efficiency determination, be it for the sample or for the external standard. As the amount of data per sample generated by the counter has increased, the presentation of the data by a lister has become very cumbersome. Considering the worst example, 22 lines of numbers are possible per sample if three external-standard channels ratios are called for in a three-channel counter. The lack of identification of each line makes this a most complex data tape to decipher. The introduction of the electric typewriter as a readout device has made the data presentation more intelligible and is a welcomed innovation. Another innovation is to employ a teletype as the standard readout device and to print either the raw data or the computed net counts per minute of each sample as it is processed as well as an external standard channels ratio. An added gimmick is the printing of the 2σ error (0.95 error) of each count per minute computation, which is merely an indication of the number of counts analyzed. For example, if a sample containing 1000 cpm net activity is counted 10 minutes in channel B, in the counts per minute mode the teletype prints B = 001000 (2.0%), the 2.0% signifying the 0.95 error of the counting rate, which means that 10,000 events have been analyzed.

The fully automated liquid scintillation counter with varying degrees of data processing capability is currently available, as manufacturers have now integrated some of their counters with minicomputers. Instant data in the form of disintegrations per minute are computed on the basis of quench curves stored in the computer. The number of quench curves and special computational routines held by the computer is simply a function of its data and memory storage capacity. Convenient as it may be, the use of a dedicated computer to process data of individual samples is of limited value for the expense involved. Furthermore, the computer is idle while the samples are counting. As of this writing, such units are designed to compute individual sample data and do not eliminate further data processing by the analyst. For example, except for highly specialized programs, the resulting data must still be averaged and then analyzed statistically. The complete analysis of the data normally performed by the analyst can be done efficiently by a large computer. To this end, the manufacturers have provided means of interfacing their respective counters to accept any standard form of data storage device for computer processing. These devices take the form of a card punch, tape punch, magnetic tape, or a direct link to a computer. A bibliography of proved software (programs) for processing data from a liquid scintillation counter is available from the manufacturers. If a large computer is not available on campus, a time-share computer accessed via telephone lines is a relatively inexpensive means of data processing by a large computer without actually investing in one.

REFERENCES

1. Crookes, W. *Chem. News* **87,** 241 (1903).
2. Regener, E. *Verh. Deut. Phys. Ges.* **10,** 78 (1908).
3. Birks, J. B. "The Theory and Practice of Scintillation Counting." Pergamon, Oxford, 1964.
4. Laustriat, F., Voltz, R., and Klein, J. *In* "The Current Status of Liquid Scintillation Counting" (E. D. Bransome, Jr., ed.), p. 13. Grune & Stratton, New York, 1970.
5. Horrocks, D. L. *In* "Organic Scintillators and Liquid Scintillation Counting" (D. L. Horrocks and C. Peng, eds.), p. 75. Academic Press, New York, 1971.
6. Kallman, H., and Furst, M. *In* "Liquid Scintillation Counting" (C. G. Bell, Jr. and F. N. Hayes, eds.), p. 3. Pergamon, Oxford, 1958.
7. Berlman, I. B. *J. Chem. Phys.* **33,** 1124 (1960).
8. Berlman, I. B. *J. Chem. Phys.* **34,** 598 (1961).
9. Birks, J. B., and Kuchela, K. N. *Proc. Phys. Soc., London* **77,** 1083 (1961).
10. Swank, R. K., and Buck, W. L. *Rev. Sci. Instrum.* **26,** 15 (1955).
11. Engstrom, R. W., Stoudenheimer, R. G., and Glover, A. M. *Nucleonics* **10,** 58 (1952).
12. Murray, R. B., and Manning, J. J. *IRE Trans. Nucl. Sci.* **7,** 80 (1960).
13. Avivi, P., Simpson, S. A., Tait, J. F., and Whitehead, J. K. *Radioisotope Conf., Proc. Int. Conf., 2nd,* 1954 Vol. 1, p. 313 (1954).
14. Bush, E. T. *Anal. Chem.* **35,** 1024 (1963).
15. Laney, B. *In* "Liquid Scintillation Counting, Current Developments" (P. E. Stanley and B. Scoggins, eds.). Academic Press, New York, 1974.

3

General Considerations of Sample Preparation

The requirement for toluene or a related substance as the base solvent for liquid scintillation counting solutions initially imposed severe limitations on the variety of samples that could be counted by this procedure. The preparation of samples tended to be an empirical affair, very much the responsibility of the individual investigator, and the literature is replete with efforts made to overcome specific problems.

Where once the objective was to devise a method that worked, nowadays the problem is usually to determine which of several methods is the most suitable for the problem in hand. Each counting situation, therefore, can usually be met in a variety of ways. Proteins, for example, labeled with carbon-14 can be combusted and the resultant radioactive carbon dioxide absorbed in a suitable trapping agent and added to a counting solution. Alternatively, the protein can be digested with a strong base and solubilized to a form suitable for counting. The choices for trapping agents, counting solutions, and solubilizers are numerous and selection is often a function of convenience.

The major considerations in choosing a method for preparing samples for liquid scintillation counting are (a) the physical and chemical characteristics of the sample to be counted (b), the isotope or isotopes contained in the sample, and (c) the anticipated level of radioactivity. The physical and chemical characteristics of the sample determine the type of counting solution required and whether conversion to a form more suitable for counting is necessary. Tritiated samples always require more attention than samples containing carbon-14 because tritium is the weakest beta emitter used and therefore the most sensitive to quenching. In addition, if samples are low in activity, a preparation that results in a markedly quenched sample may require a long or even a prohibitive amount of counting time. In contrast, for samples that contain high activity or for situations in which available counting time is not a major consideration, convenience and economy in laboratory preparation may be decisive factors in selecting a method for sample preparation. This is particularly applicable when large numbers of samples are involved.

As a general rule, the activity of the prepared sample should be such that reasonable statistical accuracy of counting can be achieved within a short counting period. In essence, this implies that the method of sample preparation should result in as little quenching as possible. This is often critical for low-activity samples, particularly tritium, but there are circumstances in which high counting efficiencies may be justifiably sacrificed for the sake of convenience. The method chosen should result in the production of a series of uniform, stable, and homogeneous samples so that counting efficiencies can be determined conveniently using, for example, the sample channels ratio for efficiency determination. If a method requiring internal standardization is chosen, further sample manipulation and additional counting time are automatically introduced.

Although there is no universal procedure applicable to all samples, a great variety of compounds can be assayed using relatively few solvent mixtures. The essential ingredients of a counting solution, in addition to the sample itself, are a solvent and a scintillator. To these are often added secondary solvents and scintillators, the former to facilitate sample preparation and the latter to give increased counting efficiencies. Before the sample itself is dealt with, the various components of the counting solution are discussed.

SCINTILLATORS

A primary scintillator should exhibit the following characteristics.

a. A high efficiency of light production when activated by radiation
b. Emission of light at a wavelength that is within the region of maximum sensitivity of the photomultiplier tube
c. Both solubility and stability under conditions imposed by the nature of the sample and the working temperature of the counter

d. No concentration quenching
e. Ready availability and low cost

Studies by Kallmann and Furst (1), Furst *et al.* (2), Hayes *et al.* (3), and Ott *et al.* (4, 5) have included an extensive survey of hundreds of compounds as possible scintillators. The major characteristic that emerges is that good scintillators are compounds containing conjugated aromatic rings combined in a linear fashion (*p*-terphenyl) rather than a fused ring system (anthracene). The Los Alamos group (3) has concluded that attempts at generalizing on the correlation of chemical structure with scintillation efficacy consistently fail because of the diverse mechanisms involved in the

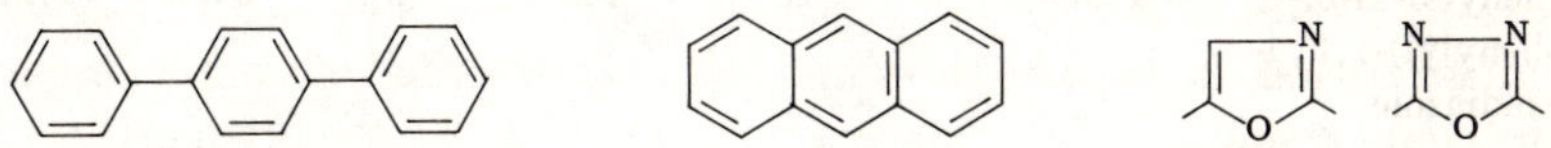

p-Terphenyl   Anthracene   Oxazole   1, 3, 4-Oxadiazole

scintillation process. However, they have arrived at four positive conclusions.

1. Linkage of four rather than three rings in a molecule yields a better scintillator, partly because of its longer wavelength spectrum.
2. Four- and five-ring compounds have sufficient solubility for scintillation applications when they incorporate heterocyclic ring systems, such as oxazole.
3. Generally beneficial ring systems are benzene, naphthalene, furan, pyrrole, oxazole, 1,3,4-oxadiazole, pyridine, indole, and benzoxazole. Less suitable are thiophene, thiazole, 1,3,4-thiazole, pyrazine, pyridazine, and benzothiazole.
4. Substituent groups causing no difficulty are methyl, methoxyl, fluoro, and chloro. Substituents reducing scintillator efficiency are bromo, iodo, nitro, and phenolic hydroxyl.

Of the many ring structures studied, oxazole and 1,3,4-oxadiazole have formed the basis of the most useful scintillators. Four primary scintillators are listed in Table 1.

p-Terphenyl was one of the first scintillators to be used, but it is not very soluble at reduced temperatures or in solutions containing water. *p*-Terphenyl can be used in ambient temperature systems and is comparatively inexpensive.

2,5-Diphenyloxazole, or PPO, is the most commonly used scintillator today. It displays excellent solubility characteristics in the usual solvents, including those used to take up water; it exhibits high scintillation efficiency

TABLE 1

Primary Scintillators

Scintillator	Abbreviation	Maximum solubility in toluene 0°C (g/liter)	Maximum solubility in toluene 20°C (g/liter)	Concentration[a] (g/liter)	Maximum fluorescence (Å)	Relative pulse height
p-Terphenyl	TP	4.0	8.6	8	3600	1.00
2,5-Diphenyloxazole	PPO	238	414	4	3650	1.01
2-Phenyl-5-(4-biphenylyl)-1,3,4-oxadiazole	PBD	10	21	9	3650	1.28
2-(4-*t*-Butylphenyl)-5-(4-biphenylyl)-1,3,4-oxadiazole	Butyl-PBD	57	119	7	3650	1.28

[a] Typical concentration in toluene for unquenched samples.

at low concentrations; it is readily available; and it is moderately priced. Its major shortcoming is that it exhibits marked concentration quenching (self-quenching) as shown in Fig. 26. This figure shows the effect on the intensity of the fluorescence induced by a gamma source in a solution containing increasing concentrations of 2,5-diphenyloxazole in xylene saturated with nitrogen, air, or oxygen. When xylene is saturated with any of these three gases, there is an optimum concentration of scintillator, different for each gas, that yields maximum fluorescence intensity. When the 2,5-diphenyloxazole concentration is further increased, quenching occurs. Oxygen has been found to be a severe quenching agent, and the quenching effect of air is directly related to its oxygen content. Figure 26 also illustrates that the optimum concentration of 2,5-diphenyloxazole for an oxygen-saturated solution results in significant quenching in a nitrogen-saturated solution, whereas it has less effect on an air-saturated solution. This emphasizes the empirical nature of determining the optimum scintillator concentration for a given counting solution because it depends on the nature of the quenching agent and the degree of quenching. There is, therefore, no specific means of determining the optimum concentrations of scintillators for a given counting situation except by experimentation.

2-Phenyl-5(4-biphenylyl)-1,3,4-oxadiazole, or PBD, is one of the most efficient scintillators known and is more resistant to concentration quenching than is 2,5-diphenyloxazole. However, about twice as much PBD is required and it is less soluble and more expensive than PPO.

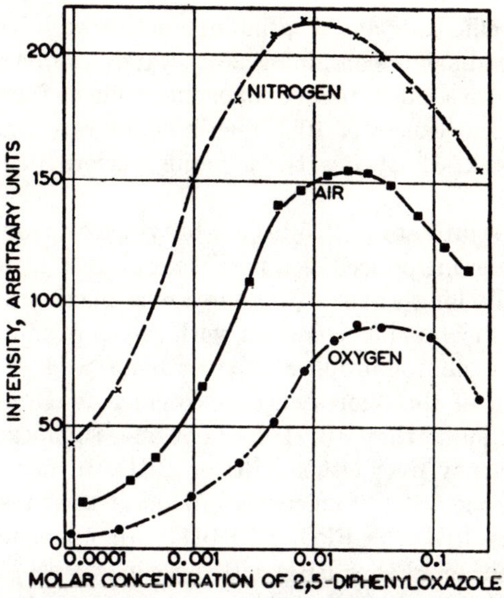

Fig. 26 Concentration and oxygen quenching of the gamma-ray induced fluorescence of 2,5-diphenyloxazole in xylene (1).

2-(4-*t*-Butylphenyl)-5-(4-biphenylyl)-1,3,4-oxadiazole, or butyl-PBD, has been synthesized to increase the solubility of PBD. This scintillator, like PBD, is required in relatively high concentrations compared to PPO, making it a relatively expensive primary scintillator. A comparison of butyl-PBD with PPO under three different quenching situations is shown in Fig. 27 (6). When quenched with either 40% ethanol or 0.5% carbon tetrachloride, butyl-PBD shows less concentration quenching and gives a

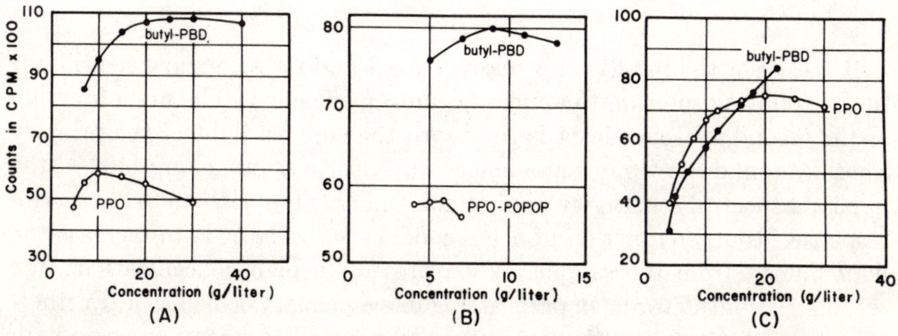

Fig. 27 Comparison of PPO and butyl-PBD as primary scintillators for counting tritiated toluene (A) in toluene containing 0.5% carbon tetrachloride, (B) in toluene containing 40% ethanol, and (C) in *p*-dioxane containing 20% water (Courtesy Searle Analytic.)

higher counting efficiency at its optimum concentration than PPO. When quenched with *p*-dioxane containing 20% water, the performance of the two scintillators are similar. In an unquenched toluene solution at 7 g/liter, butyl-PBD gives in excess of 10% higher count rate than does PPO at 5 g/liter. Figure 27 illustrates the superior performance of butyl-PBD over PPO.

A primary scintillator, 2,5-bis[5′-*t*-butylbenzoxazolyl(2′)]thiophene, "BBOT," has been introduced as a possible replacement for PPO because it has an unusually long-wavelength fluorescence maximum of 4350 Å. The longer wavelength emission coincides with the maximum photocathode sensitivity of the photomultiplier tube with an S-11 spectral response, making detection of the fluorescence more efficient. However, Bush and Hansen (7) have found that BBOT does not offer significant improvement in counting efficiency over either PPO or PBD. In fact, in unquenched toluene samples, the order of efficiencies found at their respective optimal concentrations is PBD > PPO > BBOT. In quenched samples, the combination of either PPO or PBD with a secondary scintillator, POPOP, is superior to BBOT alone. The addition of the secondary scintillator to BBOT does not improve its performance.

Efficient optical coupling between the sample and the photomultiplier tube is required to realize the optimum photon yield from the scintillator. The most commonly used light reflector is one coated with evaporated aluminum which is sometimes coupled with a plastic light pipe (8). Above wavelengths of 4300 Å, titanium oxide is a more efficient reflector than aluminum.

SECONDARY SCINTILLATORS

It has been the practice for many years to add a secondary scintillator to the counting solution to improve counting efficiency. The main function of the secondary scintillator is to absorb the emitted light of the primary scintillator and to reemit it at a longer wavelength approaching 4400 Å, the maximal spectral sensitivity of photomultiplier tubes with an S-11 spectral response. Another function of the secondary scintillator is to increase the light output from the sample. When present in high concentrations, the secondary scintillator is, in part, acting like a primary scintillator. Furthermore, under these conditions the secondary scintillator may compete with the quenching agent and reduce the quenching of the primary scintillator. The secondary scintillators are of little value when used alone because of their limited solubility. Although secondary scintillators are routinely added to most counting solutions, such primary scintillators as butyl-PBD often render their inclusion unnecessary. According to Bush and Hansen (7) a secondary scintillator should only be used when one or more of the following conditions are upheld.

1. The sample contains a compound that exhibits direct quenching of the primary scintillator.
2. The concentration of primary scintillator is high enough to produce strong self-quenching. In all samples, either quenched or unquenched, the highest counting efficiency is obtained at the maximal concentration of the secondary scintillator. Concentration quenching, if it exists for secondary scintillators, does not appear to cause a loss of counting efficiency at their solubility limit in toluene.
3. The individual counter is demonstrated to have better response at longer wavelengths.
4. The counting sample has significant absorption in the near ultraviolet.

The characteristics of four commonly used secondary scintillators are shown in Table 2. DMPOPOP has the advantage over POPOP of greater solubility but is not as efficient as POPOP and higher concentrations are required (7). Bis-MSB has been found to be slightly superior to DMPOPOP in several solvent systems (9). It also has the highest solubility of the four secondary scintillators. The newest of the secondary scintillators is PBBO and its attributes are high solubility and close spectral matching to the bialkali photomultiplier tube.

A method of systematically determining the optimum concentrations of primary and secondary scintillators has been suggested by Litle and Neary (10) and consists of preparing a ten by ten array of samples, varying primary scintillator concentrations along one axis and the secondary concentrations along the other axis, and keeping the same total volume and identical amounts of radioactivity in each sample. The counting efficiencies are determined and entered on a grid at the intersection of lines representing the corresponding primary and secondary scintillator concentrations (Fig. 28A). Appropriate contours of equal efficiencies can then be interpolated. A fixed amount of additive is then added to each sample, the samples are recounted, and the contours are redetermined, as shown for the addition of 0.2 and 0.4% nitromethane in Fig. 28B and C. These results show that increased quenching requires higher concentrations of both primary and secondary scintillators for maximum counting efficiencies to be obtained.

In the most modern liquid scintillation counters, the use of secondary scintillators is now unnecessary, and may, in fact, lower sample counting efficiency. This is because of the development of photomultiplier tubes,

TABLE 2

Secondary Scintillators

Scintillator	Abbreviation	Maximum solubility in toluene 0°C (g/liter)	Maximum solubility in toluene 20°C (g/liter)	Maximum solubility in toluene 25°C (g/liter)	Dioxane 25°C (g/liter)	Concentration[a] (g/liter)	Maximum fluorescence (Å)
1,4-Bis-2-(5-phenyloxazolyl)benzene	POPOP	0.45[b]	—	0.90	1.3	0.5	4200
		1.5[c]	2.2				4150
1,4-Bis-2-(4-methyl-5-phenyloxazolyl)benzene	DMPOPOP	1.4[b]	—	3.0	2.2	1.0	4300
		2.0[c]	3.9				4300
p-Bis(*o*-methylstyryl)benzene	bis-MSB	1.5[d]		5.5	8.5	1.0	4160
2-(4′-Biphenylyl)-6-phenylbenzoxazole	PBBO	1.6[b]		5.0	8.5	1.0	3970

[a] Typical concentrations in toluene for quenched samples with 7 g PPO per liter.
[b] Packard Instrument Company, Nuclear Chemicals and Supply Catalog, Downers Grove, Illinois (1972).
[c] J. B. Birks, "Solutes and Solvents for Liquid Scintillation Counting." Koch-Light Laboratories, Ltd., Colnbrook, Bucks, England, (1969).
[d] Pilot Chemical Company, Boston, Massachusetts.

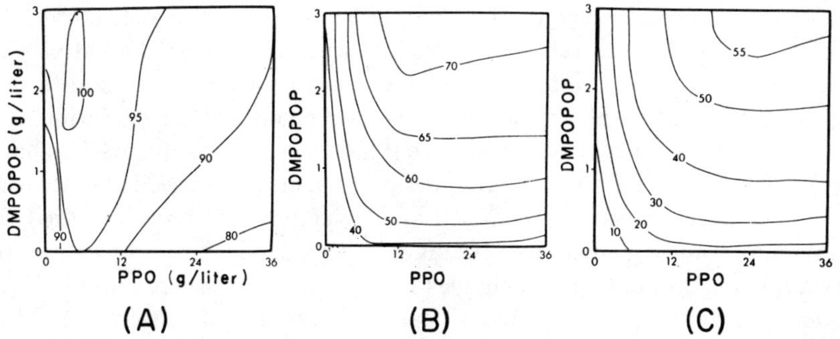

Fig. 28 Contours of equal relative counting efficiencies for the system toluene–PPO–DMPOPOP with varying amounts of PPO and DMPOPOP in g/liter. (A) Air-saturated samples containing tritiated toluene. Quenching effect of the addition of 0.2% nitromethane (B) and 0.4% nitromethane (C) to samples as in (A) (10).

specifically for use in liquid scintillation counting, that are highly efficient at the fluorescence wavelengths of the primary scintillators. In view of these recent advances in photomultiplier tube technology, it is advisable for the scientist to know which type of photomultiplier tube is used in his instrument if he is to properly formulate his counting solutions.

Scintillators are now available as scintillation grade chemicals and are commercially available as "ready to use solutions," as concentrated solutions that need only be diluted with solvent, or as prepackaged dry mixtures of scintillators that simply have to be dissolved in a predetermined volume of solvent in order to obtain the desired counting solution.

SOLVENTS

The conclusions reached in the early surveys by Hayes *et al.* (11–13), Davidson and Feigelson (14), and Furst and Kallmann (15) on solvents for use in liquid scintillation counting still serve as present day guidelines. The choice of a solvent is primarily dictated by the solubility of the scintillator in the solvent and its efficiency in transferring energy from the source of beta emission to the scintillator. It should be recalled that in the chain of energy transfers from the beta source to the emission of photons from the sample vial, the least efficient transfer of useful energy for photon production of between 5 and 10% occurs between the radioactive source and the solvent. The discovery of a more efficient solvent would be a great advance in the field. The requirements for a good solvent are (a) high solubility of the scintillator, (b) efficient energy transfer from the radioactive source, (c) transparency to the photons emitted by the scintillator, (d) ability to dissolve the radioactive sample with or without the aid of solubilizing agents, (e) ability not to freeze at the working temperature of the counter, and (f) ability to form a homogeneous counting solution.

Some of the results of two studies that have evaluated solvents are shown in Table 3. The results are different because the methods used for evaluation have not been the same. Davidson and Feigelson (14) have recorded the maximum counting efficiency obtainable with a fixed amount of scintillator by adjusting amplifier gain, whereas Hayes *et al.* (11–13) have used a fixed amount of scintillator and constant gain. The ranking of solvents according

TABLE 3

Relative Counting Efficiencies of Some Solvents

Solvent	Freezing point (°C)	Relative pulse height[a]	Efficiency[b]
p-Xylene	+13	1.12	—
m-Xylene	−47	1.09	—
Xylene (mixed isomers)	<−20	1.07	97
Phenylcyclohexane	+8	1.02	—
Toluene	−95	1.00	100
o-Xylene	−25	0.98	—
Ethylbenzene	−94	0.96	—
Triethylbenzene	−66	0.96	—
Benzene	+5	0.85	—
Methoxybenzene (Anisole)	−37	0.83	100
1,3-Dimethoxybenzene	−52	—	81
n-Heptane	−90	—	70
1,4-Dioxane	+12	—	70
1,2-Dimethoxyethane	−71	—	60
Benzyl alcohol	−15	—	38
Dimethyleneglycol diethyl ether	−44	—	32
Acetone	−94	—	12
Ethyl ether	−116	—	4
Ethyl alcohol	−114	—	0
Ethylene glycol	−13	—	0
N,*N*-Dimethylformamide	−61	—	0
2-Ethylhexanoic acid	−117	—	0

[a] All solvents contained 0.3% 2,5-diphenyloxazole (11, 12).

[b] All solvents contained 0.3% 2,5-diphenyloxazole. Toluene cpm taken as 100% (14).

to pulse heights as determined by Hayes *et al.* is more meaningful in rating solvents for scintillators. In spite of the qualitative differences shown in Table 3, the conclusion that alkylbenzenes are the best solvents has been drawn by both groups.

From their extensive studies of solvents for liquid scintillation counting, the Los Alamos group has proposed the classification of organic compounds as solvents, dilutors, and quenchers for this application (11–13). Solvents are arbitrarily classed as those compounds which give a pulse height of at least 80% of that achieved in toluene with a given concentration of scintillator. Dilutors are those compounds which cause a gradual depression of pulse height proportional to their concentration. Quenchers are those compounds which cause a marked drop in pulse height at low concentrations. The effect of a typical dilutor, 1,4-dioxane, and a typical quencher, acetone, on the pulse height of a toluene solution containing 3 g/liter of 2,5-diphenyloxazole are shown in Fig. 29 A and B. Organic compounds classed as dilutors include aliphatic alcohols, esters, ethers, nitriles, saturated hydrocarbons, olefins, fluoro- and chloroalkanes, and methylborate. Dilutors are especially valuable as solubilizing agents for various types of compounds that are not very soluble in toluene alone and for use as an antifreeze in dioxane-containing counting solutions. Organic compounds classed as quenchers include aliphatic acids, aldehydes, amines, ketones, and thiophenes, as well as chloro- and bromobenzenes and nitrogen-containing heterocyclic compounds.

Toluene has remained the solvent of choice from the point of view of relatively efficient energy transfer, cost, and availability. This has prompted

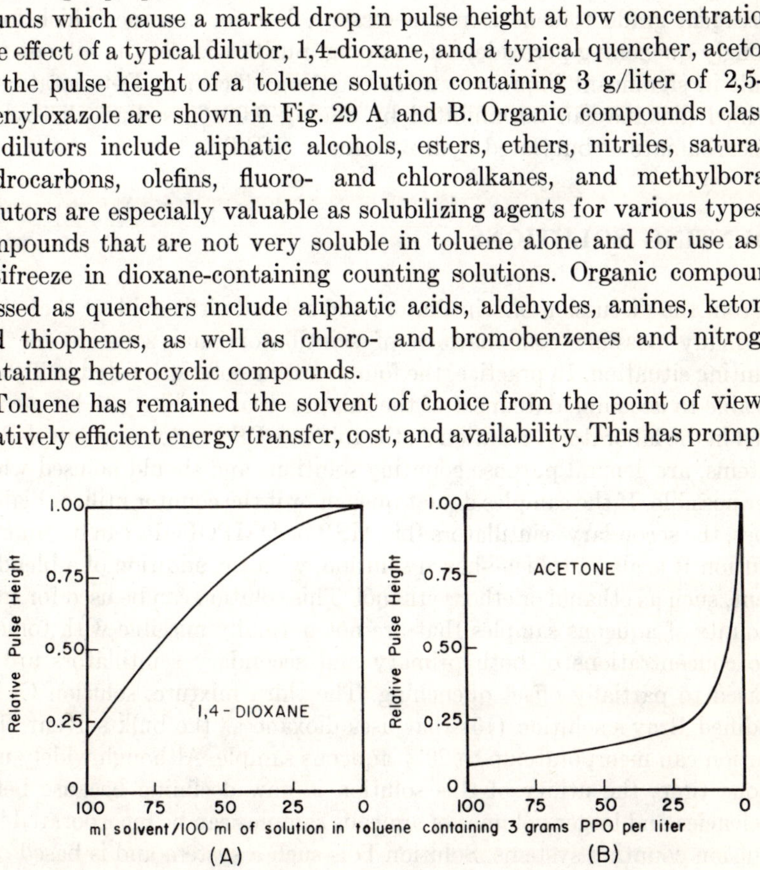

Fig. 29 Effect of varying amounts of (A) a typical solvent dilutor, 1,4-dioxane, and (B) a typical solvent quencher, acetone, in solution with toluene containing 3 g of 2,5-diphenyloxazole per liter of solution on the relative scintillation efficiencies compared to that in toluene alone (12).

the appearance of "scintillation-grade" toluene but, as far as can be determined, the less expensive "reagent-grade" toluene is quite satisfactory for use in liquid scintillation counting. Of the other solvents listed, 1,4-dioxane is of particular interest because it is miscible with water and until recently has been a widely used solvent for incorporating aqueous samples. Although dioxane freezes at 12°C, rendering it unsuitable when the counter is operated at reduced temperatures, this obstacle has been overcome either by adding secondary agents to act as an antifreeze or by counting at room temperature.

The solvent comprises the bulk of the sample and, therefore, presents the greatest potential danger for containing quenching impurities. Purification of reagent-grade toluene is not necessary, but for dioxane, which forms strongly quenching peroxides on standing, purification of the solvent may result in significant improvement in counting efficiency. Deterioration of purified dioxane can be inhibited by adding 0.001% sodium diethyl dithiocarbamate or butylated hydroxytoluene (BHT). .

COUNTING SOLUTIONS

From the discussion of scintillators and solvents it is evident that there are many possible combinations of scintillators and solvents for any counting situation. In practice, the four counting solutions listed in Table 4 are able to accommodate most of the samples that are likely to be encountered in the laboratory. Solution A, the toluene–PPO or toluene–butyl-PBD systems, are general purpose counting solutions and should be used whenever possible. If the samples do not quench or if the counter utilizes bialkali tubes, the secondary scintillators (bis-MSB or DMPOPOP) can be omitted. Solution B is also a toluene-based solution, with the addition of a blending agent, such as ethanol or ethoxyethanol. This solution can be used for small amounts of aqueous samples that are not normally miscible with toluene. The concentrations of both primary and secondary scintillators are increased to partially offset quenching. The third mixture, solution C, is a modified Bray's solution (16) that uses dioxane as the bulk solvent. This solution can incorporate up to 20% aqueous sample. Although widely used at one time, the utility of this solution is now declining because better efficiencies and larger volumes of aqueous samples can be incorporated into emulsion counting systems. Solution D is such a system and is based on a Triton X-100–toluene combination. The major problem with this system is the batch-to-batch variation in Triton X-100 and each new preparation of counting solution, therefore, must be standardized. It should also be stressed that although the formulations of the various solutions listed in

TABLE 4

Composition of Some Standard Counting Solutions

Solution	Composition				Uses
	Primary scintillator	Secondary scintillator	Additives	Solvent (to 1 liter)	
A	PPO (5 g) or butyl-PBD (10 g)[a]	bis-MSB (0.5 g) or DMPOPOP (0.25 g)	— —	Toluene	All samples soluble in toluene; insoluble samples adsorbed on inert supports
B	PPO (8 g) or butyl-PBD (10 g)	bis-MSB (1 g) or DMPOPOP (0.5 g)	Ethanol or 2-ethoxyethanol (300 ml)	Toluene	Aqueous samples up to 3%
C	PPO (8 g) or butyl-PBD (10 g)	bis-MSB (1 g) or DMPOPOP (0.5 g)	Naphthalene (150 g), ethyleneglycol (20 ml), and 2-ethoxyethanol (100 ml)	Dioxane	Aqueous samples up to 20%
D	PPO (5 g) or butyl-PBD (10 g)	bis-MSB (0.5 g) or DMPOPOP (0.5 g)	Triton X-100 (333 ml)	Toluene	Aqueous samples below 10% and above 20% to a maximum of about 40%

[a] No secondary scintillator used with butyl-PBD.

Table 4 are adequate for most counting situations, the concentration of the various components may not be optimal for a given type of sample. If maximum counting efficiency is required, therefore, the formulations of these counting solutions may have to be modified.

The relationship between the volume of the counting solution and counting efficiency is shown in Fig. 30. This is obtained by preparing a series of toluene solutions of varying volumes but containing a fixed amount of tritium as labeled toluene and the scintillator, butyl-PBD, at a concentration of 8 g/liter. The samples were counted in a standard, low-potassium glass, 22-ml capacity counting vial and in a so-called "minivial," which is a low-potassium glass vial with a 12-ml capacity. The minivials are placed inside a plastic vial that functions simply as a reusable minivial holder. Figure 30 shows that the optimum volume for the standard vials is about 12 ml, whereas that for the minivials is about 6 ml. Reasonable efficiency can be had using only 3 ml of counting solution in a minivial or 4 ml in a standard vial. The optimum volume for any counter can be altered, within limits, by adjusting the sample elevator stop on most counters, although most instruments are factory adjusted so that 12- to 15-ml samples are optimal in terms of sample–photomultiplier tube geometry. The advantages of using smaller volumes of solution are (a) a reduction in tritium background of roughly 0.5 cpm/ml of counting solution in a wide window and (b) a reduction in the cost of sample preparation because the solvent and scintillator costs are reduced by a half and the cost of minivials is only 40% that of the standard low-potassium vials at current prices. If the sample itself contains a quencher, then larger volumes of the counting solution may be desirable to reduce the quenching by dilution.

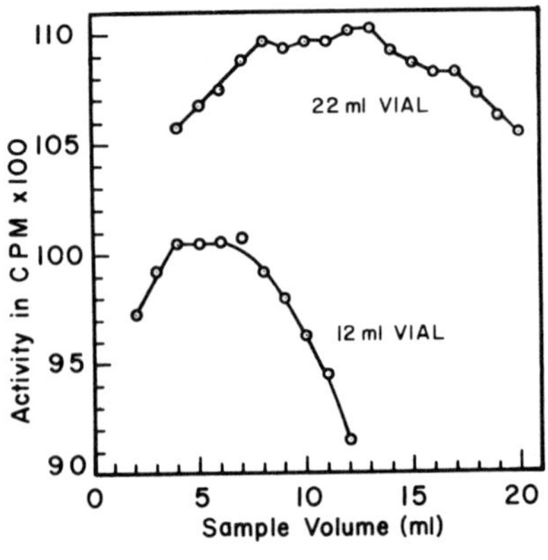

Fig. 30 Relationship of counting solution volume to counting efficiency of a standard tritium-labeled toluene solution containing 8 g butyl-PBD per liter of toluene.

COUNTING VIALS

The standard counting vials used for the past decade have been the 20-ml glass vials made of borosilicate containing minimal amounts of potassium-40 and other radioactive elements. These vials now exhibit a high degree of uniformity and low-background performance. However, there are available less expensive glass vials that are suitable for samples that can tolerate variations in background of between 2 and 5 cpm. A most welcomed development has been the introduction of the low-potassium, 12-ml glass vial, which has proved to be a uniform product of equal quality to the larger glass vial but at about 40% of its cost. A glass container is generally preferred over plastic vials because it can be treated as a chemical vessel. In the very early reports of liquid scintillation counting, quartz vials have occasionally been used when the lowest background counts have been sought. Quartz vials do reduce the tritium background significantly when compared to glass vials (20–30%) but the carbon-14 background is usually the same. The high cost of quartz vials has prevented their general use.

Plastic vials have the best low-background characteristics. In this category nylon vials are superior and are relatively impervious to toluene. However, nylon vials swell in water and are not useful for aqueous samples. In fact, this property of nylon is used to detect moisture in humidity determinations.

Vials made of high-density polyethylene are the most widely used but polyethylene is slightly permeable to organic solvents and this causes the vials to swell. The first medium-density polyethylene vials lost as much as 100 mg of toluene per vial per day, but this has now been reduced to about 35 mg per day. Nevertheless, it is inadvisable to store samples in polyethylene vials. Recently, a disposable 6-ml capacity polyethylene vial has been introduced. This vial is transported in a reusable clear plastic carrier that fits into the sample belt of automatic counters. An interesting observation with plastic vials is that higher counting efficiencies are observed than with glass vials. This phenomenon is caused by improved light collection resulting from the lack of light trapping within the plastic vial. Benson and Maute (17) have observed increased tritium counting efficiency when glass sample vials are frosted with condensation in the cold sample chamber. The efficiency is reduced and remains constant when the condensation disappears and this phenomenon has been correctly attributed to more efficient light collection rather than to induced phosphorescence of the sample or vial. A detailed study of internal light reflection within the glass sample vial has been carried out by Gordon and Curtis (18). They have found that increased counting efficiencies are observed when the samples contain optically diffusing white materials, such as a piece of white fabric, or when the exterior vial surfaces are sandblasted. These findings support their theoretical calculations that the increased counting efficiencies are caused by a reduction in the amount of light lost by internal reflection. They have also found that the increased counting efficiency can be monitored satisfactorily with an internal standard or by the channels ratio technique, using either the sample count ratios or the external standard ratios.

A drawback of plastic vials is their tendency to exhibit phosphorescence, which can persist for days. When this occurs, the samples must be dark adapted in the counter before they are counted. The major disadvantage of plastic vials, as already stated, is the permeability of organic solvents, a feature that can cause a significant change in the counting characteristics of the sample.

For smaller amounts of radioactive material, adsorption on the walls of the vial can be a major problem. This may occur with either plastic or glass vials, depending on the nature of the material. If surface adsorption occurs, it can be overcome by the prior addition into the vial or the counting solution of a small amount of silica gel (19), which is composed of submicron particles and provides over a 10,000-fold greater surface area for adsorption and helps maintain 4π geometry for efficient counting. Another method of preventing surface adsorption on the glass wall is to siliconize the vial with dimethyldichlorosilane dissolved in toluene (20). A 20–40% decrease in the count rate of ^{32}P- and ^{45}Ca-containing samples over a 24-hour period has been prevented by this treatment. However, another group (21) has found that coating glassware with silicone-containing hydrophobic agents does not prevent the adsorption of certain radioactive ions and has concluded that, in general, glass vessels adsorb less than those of polypropylene.

SOURCES OF BACKGROUND COUNTS

The background count rate is always subtracted from the observed sample count rate to obtain the net sample count. When the sample count rate is high compared to the background activity, the background can be ignored but in the analysis of low-activity samples the sensitivity of the determination becomes a function of the background count rate. Knowledge of the sources of the background count rate for a given counter can help, therefore, in achieving low-background conditions when they are required.

In a liquid scintillation counter, one of the sources of background is cosmic rays, which constitute the external source of background counts. Cosmic rays are high-energy particles that interact with matter to produce mesons, electrons, gamma rays, and X rays which, in turn, irradiate the counter from all directions. Cosmic rays are responsible for Cerenkov radiation, which can occur within the liquid sample. Cerenkov radiation results from the passage of a particle through the medium at a speed greater than that of light through that medium. The Cerenkov radiation energy (photon production) is a function of the refractive index of the medium and the energy of the beta particle (22). The counter is protected from cosmic rays by shielding with a high-density material but it is not possible to completely eliminate the background counts. It should be pointed out that counters located at high altitude, such as those in Denver, Colorado, which is a mile above sea level, have higher backgrounds than those at sea level because of the increased exposure to cosmic radiation, which has not been attenuated by air. The background events from cosmic radiation are of relatively high energy and can be largely excluded during pulse height analysis for tritium and carbon-14 content.

A major source of background counts is the naturally occurring radioisotopes contained in the materials used in the manufacture of the counter, the sample container, and the sample itself. The sources of background for a given counter can be determined, experimentally, by first counting without any sample in the counting chamber, then with an empty vial, followed by the vial filled only with the counting solution without any radioactive sample. The results of such a determination using a controlled-temperature counter equipped with bialkali photomultiplier tubes are shown in Table 5. Four types of vials are used in groups of ten. The empty vials are counted twice for 10 minutes each time, then filled with 5 or 10 ml of toluene containing 8 g of butyl-PBD per liter of solution, and recounted once for 10 minutes. The counts shown for the vials plus solvent represent the total background counts of the system, including those from the counting chamber. In the tritium channel, the importance of the sample container with respect to background is striking in that the use of a plastic vial can reduce it from 27 cpm for glass to 12 cpm for polyethylene in the regular 20- to 22-ml size containers. In the minivials, a similar reduction from 18 to 8 cpm is obtained in tritium counting. However, the carbon-14 backgrounds for all types of vials are very similar.

In the carbon-14 channel, the major source of background events is the electronic noise of the photomultiplier tube, consisting of the spontaneous

TABLE 5

Sources of Background Counts in Air-Quenched Samples of Toluene Containing 0.8% Butyl-PBD

Counting conditions[a]	Tritium channel	Carbon-14 channel
Tritium counting efficiency (%)	60	0.05
Carbon-14 counting efficiency (%)	20	63
Total background of system (cpm ± S.D.)		
20-ml polyethylene vial + 10-ml of counting solution[b]	11.76 ± 0.91	18.40 ± 1.53
22-ml low-K glass vial + 10 ml of counting solution	27.19 ± 1.91	18.57 ± 1.72
12-ml low-K glass vial + 5 ml of counting solution in polyethylene holder	17.70 ± 1.25	16.33 ± 1.82
7-ml polyethylene vial + 5 ml of counting solution in plastic holder	8.94 ± 0.95	15.34 ± 0.77
Distribution of background counts (cpm ± S.D.)		
empty counting chamber[c]	4.42 ± 0.46	15.41 ± 0.90
empty 20-ml polyethylene vial	2.93	1.57
empty 22-ml low-K glass vial	18.00	0.39
empty 12-ml low-K glass vial in polyethylene holder	9.80	1.89
empty 7-ml polyethylene vial in plastic holder	3.12	0.42
10 ml of counting solution	4.49	2.09

[a] All samples consisted of ten replicates. Empty vials were counted 10 minutes, twice. Vials with counting solution were counted once for 10 minutes. The distributions of background counts were determined by difference. The count attributed to 10 ml of counting solution was the average of values found by difference for polyethylene and glass vials.

[b] Toluene containing 8 g butyl-PBD per liter.

[c] Counted for 220 minutes (11 × 20 minutes).

thermionic emission of electrons, and the light flashes that occur within the tube as a result of recombination of ions produced either by the ionization of the residual gas or by the bombardment of sensitive surfaces by the accelerated electrons (23). This light is seen by the other photomultiplier tube, resulting in a coincident pulse. From Table 5 it can be seen that more than 80% of the carbon-14 background count originates in the empty counting chamber. The sample container and the counting solution are minor contributors to background in this particular carbon-14 channel. It should also be noted that organic solvents prepared from fossil sources have much lower natural content of carbon-14 because of isotope decay and, therefore, lower background. For example, *p*-cymene prepared synthetically from fossil sources contains about 1.2 dpm/ml, whereas that prepared from natural oils of plants and trees contains about 12 dpm/ml (24). Toluene is prepared commercially almost exclusively from fossil sources.

Phosphorescence is a troublesome source of background events. Birks (25) has pointed out that there is an unfortunate ambiguity in terminology with respect to phosphorescence. For our discussion, "phosphorescence" is defined as the delayed and persistent emission of light following excitation of matter, as differentiated from fluorescence, which is defined as the instantaneous emission of light following excitation. Duration of fluorescence is of the order of 10 nseconds, whereas phosphorescence can vary from ten mseconds to days. Phosphorescence can be further differentiated on the basis of its decay pattern, as either the fast or slow component, by arbitrarily defining the slow component of phosphorescence as that which requires more than a few minutes to decay. Glass vials can be activated by the ultraviolet in sunlight, by conventional fluorescent lighting, or by daylight transmitted through window glass (14, 25). In a scintillation counter, the apparent decay time of this type of phosphorescence is of the order of 2–3 minutes but strongly photoactivated samples can remain active for several days (14). In view of the inherent threat of photoactivation of samples, it is prudent to avoid exposing the sample vials unnecessarily to direct sunlight and locating counters close to bright fluorescent lighting. The latter can be a hazard with table top counters located beneath a fluorescent light that directly illuminates the samples when the sample belt is exposed for loading. Phosphorescence can be eliminated by dark adapting the samples. Fortunately, the photomultiplier tubes are protected by light shields so that photoactivation of the glass envelope is not a problem.

Chemiluminescence falls into the same category as phosphorescence. Both are random, single-photon events that if sufficiently intense result in a pileup of single photons at the photocathode of the photomultiplier tube at a rate that permits some photons to avoid exclusion by the coincidence circuit. Apparent coincident counts are therefore generated with the consequent increase in background rate. In chemiluminescence, photons are generated as a result of chemical interaction of the sample components. For example, when proteins are digested in alkali, the development of chemiluminescence is not uncommon (26, 27). It can be minimized by the addition of a small amount of acid. The intensity and duration of chemiluminescence are temperature dependent and chemiluminescence decays significantly faster at lower temperature than at room temperature. Chemiluminescence that decays to background levels in 2–3 hours in a cooled counter may still persist after 24 hours in an ambient counter. However, chemiluminescence can sometimes be severe and persistent and only overcome by allowing it to decay in the dark over a period of several days (26). This has been reported by Gower and Bransome (28) for protein-containing acrylamide gel fractions that have been dissolved in Bio Solv. Their samples are pretreated with catalase to remove peroxide, which has been thought to be the initiator of chemiluminescence. Even then, Gower and Bransome recommend dark adaptation of the samples for 2 or 3 days before they are actually counted.

Plastic vials can be a source of inconsistent background events because of their tendency to become charged with static electricity. This creates two problems: First, when the static electricity leaks off the vial slowly in the counting chamber, the light emitted by the discharge causes spurious counts to be registered; second, the statically charged vials attract dirt that can be carried down into and accumulate within the counting chamber, resulting in a slow apparent drift downward in counting efficiency as a result of deteriorating optics. Plastic vials should therefore be wiped clean with a moist cloth to remove dirt and to discharge any static electricity.

An obvious, though sometimes forgotten, source of background is the location of the counter near such high-energy sources as gamma irradiators, neutron generators, and similar instruments. A liquid scintillation counter should be isolated from these installations.

COUNTING IN SOLUTION

Aqueous Solutions

A variety of counting solutions have been evolved for counting aqueous solutions containing different isotopes. The simplest case is that of tritiated water, which is important as it is often counted as such after combustion of organic substances. Small quantities, less than 5% of the final volume, are usually incorporated into a toluene-based counting solution to which has been added a more polar solvent, such as ethanol or ethoxyethanol. Larger amounts of aqueous samples are better counted in an emulsion system or in a dioxane-based solution. Although water is a strong quenching agent this quenching can be largely overcome by adding naphthalene to the counting mixture. Naphthalene acts as an intermediate solvent, transferring the primary absorbed energy to the scintillator more effectively than does the primary solvent. The effectiveness of naphthalene, however, varies with the nature of the quenching agent. For example, the addition of naphthalene (100 g/liter) to a toluene scintillator triples the counting efficiency of solutions quenched with methyl benzoate. In contrast, naphthalene causes a decrease in the counting efficiency when the same liquid scintillator is quenched with benzoic acid (29).

Whenever possible the ratio of the volume of water to the volume of the counting solution should be made optimal to improve the performance. To this end the term "figure of merit" has been introduced. The figure of merit is the product of the water volume (*V*) and the counting efficiency (*E*). By plotting the figure of merit (*EV*) against the water volume (*V*) the optimum volume of water can be deduced from the curve. Such a curve for a homogeneous solution consisting of water–methanol–toluene has been published by Mahin and Lofberg (30). Figure 31 shows the proportions of toluene and methanol producing a single-phase system as a function of water content at a constant final volume of 17 ml. Figure 32A shows the figure of merit for the toluene–methanol–water system as a function of water content. Figure 32B shows a similar plot for Bray's solution (16) for comparison purposes. Bray's solution, which is basically a dioxane-based solution containing naphthalene, methanol, ethyleneglycol, PPO, and POPOP, has been a very popular counting solution for large amounts of water but is now being replaced by the emulsion systems.

White (31) has assessed the relative efficiencies and costs of some liquid scintillation mixtures for aqueous tritium samples. In toluene-based solutions ethoxyethanol has been found superior to ethanol as a blending agent.

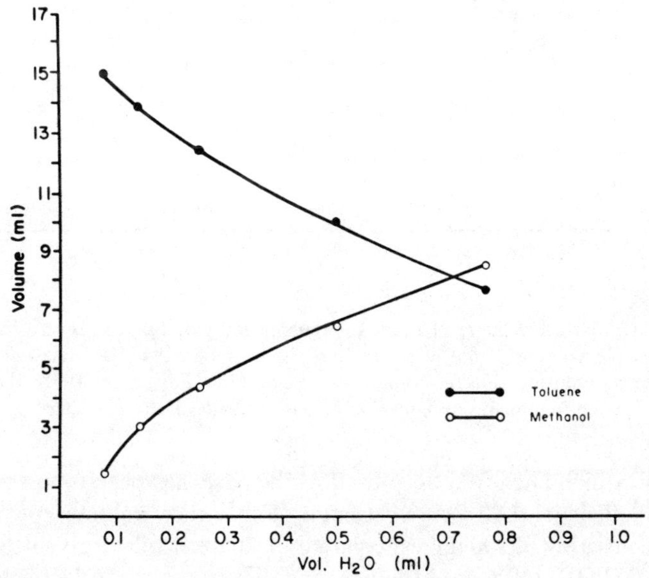

Fig. 31 Proportions of toluene (●) and methanol (○) producing a single-phase system as a function of water content. The final volume is 17 ml (30).

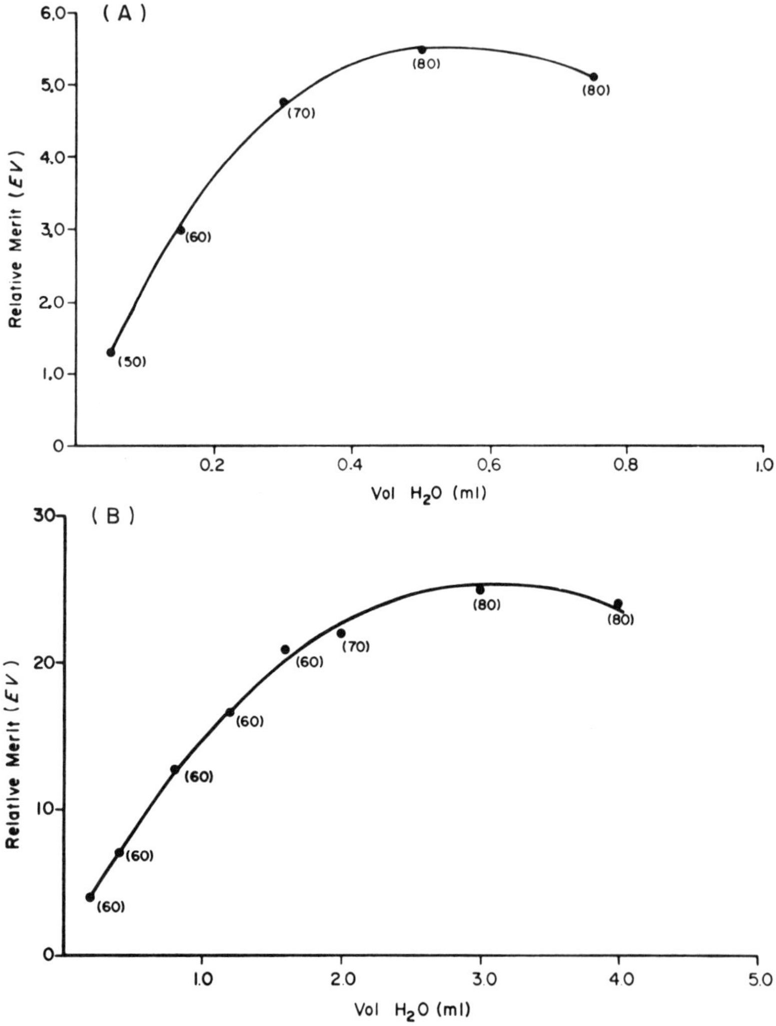

Fig 32 (A) Relative figure of merit for toluene–methanol mixtures as a function of aqueous sample volume. Toluene containing 6 g/liter of PPO. (B) Relative figure of merit of Bray's Solution as a function of sample volume. For (A) and (B), the counter gain was adjusted for maximum counting efficiency for each sample (30).

Methanol, which has often been used in the past, has been found considerably less efficient than ethoxyethanol. For larger volumes of water a solution consisting of 100 ml ethoxyethanol, 20 ml ethylene glycol, 8 g PPO, 600 mg DMPOP, 150 g naphthalene, and dioxane to make 1 liter has been found the best compromise with regard to cost, water capacity, and counting efficiency. Commercial-grade toluene and naphthalene have also been found to give counting efficiencies as good as the more expensive scintillation-grade reagents.

Solubilizing Agents

One of the major limitations in the early development of scintillation counting as a practical aid to research has been the use of aromatic solvents. Many biological samples are normally insoluble in toluene but by treating them with a solubilizing agent most types of samples can then be incorporated into a toluene-based counting solution. The more versatile of the solubilizing agents are Hyamine, NCS, Bio Solv, Soluene, and Protosol. All these agents, however, interact with butyl-PBD to produce a colored solution with consequent reduction in counting efficiency (32, 33). PPO, therefore, should be used as the primary scintillator in the counting solution.

Hyamine

Hyamine hydroxide 10-X, *p*-(diisobutylcresoxyethoxyethyl)dimethylbenzylammonium hydroxide, was the first solubilizing agent to be used with any success and is still widely used. Gradually, however, it is being replaced as superior agents become available. It was first used as a trapping agent for $^{14}CO_2$, for the carbamate is soluble in toluene, and it is still used for this purpose. Hyamine is supplied in a methanol solution and this can be used directly. Up to 20 mg of protein can be solubilized directly in Hyamine, although heating is usually necessary. Hyamine can also be used for urine, serum, plasma, and amino acids. The sample should be solubilized with Hyamine before the counting solution is added. When diluted with a toluene scintillator, it is not a very effective solubilizing agent.

There are two major limitations to the use of Hyamine. First, it has a marked quenching effect and, therefore, minimal amounts should be used. Second, many samples prepared in Hyamine exhibit chemiluminescence. This results in spurious, high count rates that are independent of the sample activity. This can often be eliminated by dark adaptation of the samples or by the addition of acid to neutralize the basicity of the counting solution. Chemiluminescence is enhanced when impure dioxane solutions are used and decay is prolonged at ambient temperatures.

NCS

NCS is a quaternary ammonium base, soluble in toluene and furnished in 0.6 *N* toluene solution. It is comprised of a mixture of bases in the molecular weight range 250–600 produced from a commercial mixture of chlorides of the type formula $R_2R_2{}^1NCl$, where R is methyl and R^1 is a straight chain varying from C_6 to C_{20} with an average of about 12 carbon atoms. NCS is more soluble in toluene and less quenching than Hyamine. It is used in a similar manner and for the same purposes as Hyamine.

Bio Solv

These solubilizing agents have been available formerly under the name of Colosolve. BS-1 is a general purpose agent that has been largely superseded by BS-3 for counting all kinds of aqueous solutions, including urine and blood plasma. The solubilizer is simply added to the sample, followed by the counting solution, and shaken. BS-2 is an acid solubilizer used to solubilize alkaline tissue digests or strongly alkaline aqueous solutions. About 50 mg of tissue is placed in a counting vial and digested at 80°C for about an hour in 2 *N* NaOH. After cooling, enough BS-2 is added to bring the pH to about 7. A toluene–PPO counting solution is then added. The background count for these solubilizing agents is low and phosphorescence and chemiluminescence are not a major problem.

Soluene

In contrast to NCS, Soluene is a single compound formed from a quaternary ammonium chloride. It is supplied in 0.5 *N* toluene and is stable on storage. Soluene is used in a similar manner to, and for the same purposes as, Hyamine and NCS.

Protosol

Protosol is a 0.5 *M* quaternary ammonium hydroxide solution in a proprietary solvent designed to dissolve, solubilize, and incorporate many types of biological tissue in counting solutions normally used in liquid scintillation counting, such as toluene, xylene, and dioxane. Its intended use is similar to that for Hyamine, NCS and Soluene.

Other products are sure to be introduced into the marketplace as the use of tissue solubilizers become more popular but the choice between these agents is a matter determined largely by experiment. A common advantage of these compounds is their versatility. They permit biological specimens as diverse as tissues, proteins, blood, amino acids, plasma, and urine, as well as aqueous samples, to be incorporated directly into a toluene cocktail. They are not, however, without their shortcomings. To differing degrees they are all quenching agents. Furthermore, many samples must receive additional treatment to remove color. This is done by bleaching the sample with a decolorizing agent, such as benzoyl peroxide. Hydrogen peroxide can also be used but it is more quenching than benzoyl peroxide. All quaternary ammonium bases attack the tinfoil lining the caps of glass scintillation vials, resulting in marked quenching of the sample. When substantial quantities of these agents are used, the lining should be replaced by a Teflon insert.

EMULSION COUNTING

A widely used emulsifying agent is the surface-active compound Triton X-100 (34–40). A system containing a toluene cocktail and Triton X-100 in the proportions 2:1 (v/v) is the most satisfactory. This system is capable of incorporating up to 15% of aqueous sample and gives a clear solution and good counting efficiencies for carbon-14.

The physical nature of the emulsion changes as the aqueous concentration is increased. When the concentration is below 15% the emulsion is clear but as the concentration is increased, two distinct phases appear. Emulsions in this form are unstable and of little value for counting. When the concentration of the aqueous phase is further increased a stable emulsion results. This may be clear or opaque but the counting efficiencies are lower than with the clear solutions obtained at lower aqueous concentrations. For measuring samples labeled with tritium the experimental conditions are more critical than for carbon-14 (35, 37, 38). Careful manipulation of the ratios of toluene cocktail:Triton X-100:aqueous sample is essential if reliable and reproducible results are to be obtained.

The demand for accommodating various types of aqueous samples in a counting solution for liquid scintillation counting has prompted the development of commercial products to satisfy this need. Such products include Instagel, Aquasol, PCS, Ready-Solv 1, Multisol, and Handifluor. These products have been formulated as either toluene- or xylene-based emulsifiers that can accommodate up to 50% water. The physical characteristics of the resulting sample depend on the amount of water contained in the sample. A typical phase diagram and typical performance data of Aquasol are shown in Fig. 33 and Table 6.

The great attraction of emulsion counting is its ability to accommodate relatively large amounts of water and salt solutions for counting at reasonable sample counting efficiency. However, the evidence is not at all clear that this is a simple technique in which the aqueous sample is simply shaken up with the emulsifier and counted. Emulsion counting appears to require rather strict adherence to a defined regimen if reliable data are to be generated. The clear phase of the Triton X-100 emulsion has been called a

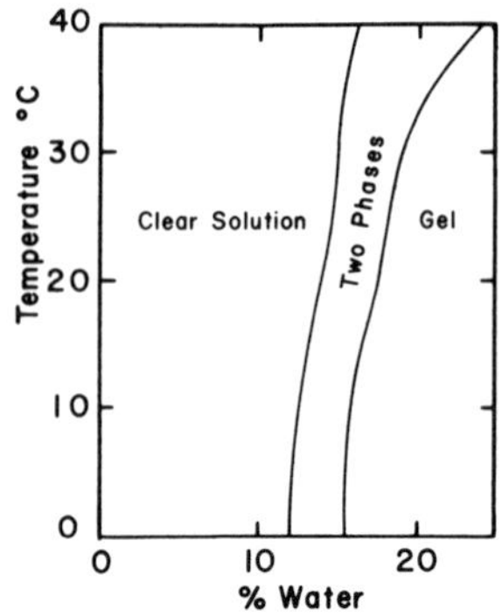

Fig. 33 The phase diagram of a typical commercial emulsifier, Aquasol. The physical and optical characteristics of the solution change with increasing aqueous content, ranging from a clear liquid to an opaque, stiff gel. Between these two states, there is a narrow range of water content at which phase separation occurs. The addition of water eliminates this phasing and results in the formation of a homogeneous gel of long-term stability. (Courtesy New England Nuclear Corp.)

true solution by Benson (37) and Turner (34), whereas van der Laarse (38) classifies all phases as suspensions of water in toluene–Triton X-100. From the phase diagram for Aquasol shown in Fig. 33 it can be seen that the phases are temperature dependent. Benson (37) and Williams and Florkowski (39) have studied the effect of temperature on the counting efficiency of a Triton X-100–toluene–water emulsion and have found that the counting efficiency increases linearly about 10% with decreasing temperature from about 17°C to a maximum between 4° and 0°C. Lower temperatures do not increase the counting efficiency further. These authors (37, 39) have also found that the emulsions give reproducible results only if they are first heated to 40°C, allowed to cool without shaking, and kept at 4°C for 2–4 hours. Emulsions prepared simply by shaking vigorously at room temperature, although clear in appearance, have been found unstable as judged by lower and more inconsistent counting data. A tenable explanation for these observations may be that heating to 40°C excludes trapped oxygen from the emulsion, and by cooling without mixing air adsorption can be minimized, thus reducing oxygen quenching. When samples are vigorously shaken, an increased amount of oxygen can be trapped in the emulsion and, on cooling, become more soluble and dissolve in the emulsion, thus increasing quenching. The sensitivity of emulsion counting efficiency to temperature and to the method of preparation dictate that these factors be strictly controlled if meaningful data are to be generated.

In Fig. 34 is shown a plot of the counting efficiency of tritiated water or carbon-14 glucose when increasing amounts of water are added to a toluene–

TABLE 6

Performance Data for a Commercial Emulsifier, Aquasol[a]

Sample type	Sample volume (ml)	Aquasol volume (ml)	Percent of sample	^{3}H Counting efficiency (% range)	Figure of merit[b]	Physical characteristics: Ambient	Physical characteristics: 5°C
	0.0	15	0.0	30–50	0	Clear	Clear
H_2O	0.5	15	3.2	25–45	106	Clear	Clear
H_2O	3.5	15	18.9	20–40	575	Cloudy gel	Cloudy gel
H_2O	5.0	15	25.0	20–40	700	Hazy gel	Hazy gel
0.5 *M* HCl	0.5	15	3.2	15–35	99	Slight haze	Clear
0.5 *M* HCl	1.5	15	9.1	20–45	241	Clear	Clear
0.5 *M* HCl	3.5	15	18.9	20–40	493	Stiff, cloudy gel	Stiff, cloudy gel
0.9 *M* NaCl	0.5	15	3.2	15–30	105	Cloudy	Clear
0.9 *M* NaCl	1.5	15	9.1	25–40	260	Clear	Clear
0.9 *M* NaCl	3.5	15	18.9	20–40	567	Cloudy	Cloudy gel
0.5 *M* NaOH	0.5	15	3.2	20–35	96	Cloudy	Hazy
0.5 *M* NaOH	1.5	15	9.1	25–40	242	Clear	Clear
0.5 *M* NaOH	3.5	15	18.9	20–35	516	Clear	Cloudy gel
30% Sucrose	1.5	15	9.1	20–35	270	Cloudy	Clear
30% Sucrose	3.0	15	16.7	20–35	448	Clear	Clear
30% Sucrose	6.0	15	28.6	15–30	712	Clear	Cloudy gel

Sample type	Sample volume (ml)	Aquasol volume (ml)	Percent of sample	^{14}C Counting efficiency (% range)	Figure of merit[b]	Physical characteristics: Ambient	Physical characteristics: 5°C
H_2O	0.5	15	3.2	70–90	281	Clear	Clear
H_2O	5.0	15	25.0	70–90	2133	Stiff, hazy gel	Stiff, hazy gel

[a] New England Nuclear Corp., Boston, Massachusetts.

[b] Figure of merit = percent of sample × efficiency. Six samples at each point. Internal standard = 1.19×10^5 dpm.

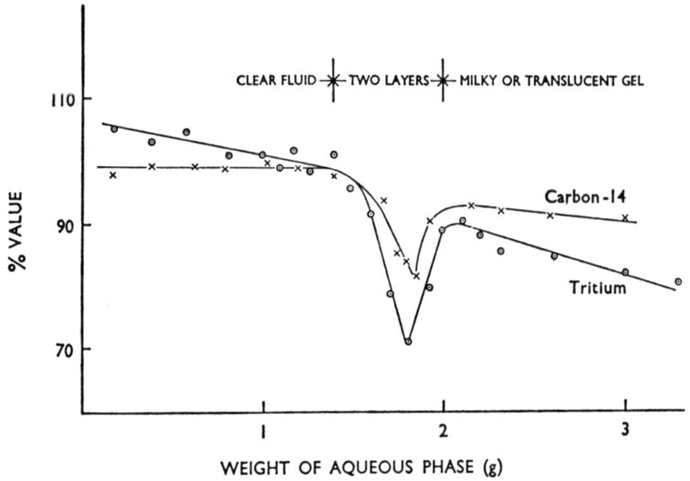

Fig. 34 Emulsion counting of [U-^{14}C]D-glucose and tritiated water using toluene–Triton X-100, 2:1 (v/v) (34).

Triton X-100 (2:1, v/v) emulsion (34). As the concentration of water is increased, the counting efficiency of the tritium in solution slowly decreases because, in this instance, water is acting as a quenching agent. When the first phase change occurs, phase separation into an aqueous and organic layer is observed and there is a dramatic fall in counting efficiency because the tritium stays in the aqueous phase, whereas the scintillator remains dissolved in the organic phase. Only the tritium at the interface between the organic and aqueous layers has sufficient energy (mean free path) to penetrate the organic layer and interact with the scintillator. As more water is added, an emulsion is formed that consists of submicron water particles dispersed in the organic phase. The size of the water particles are less than the mean free path of the average tritium beta particle and therefore the beta particles emitted by tritium can interact efficiently with the scintillator with a consequent jump in counting efficiency. The opacity of the emulsion also helps to prevent internal light trapping, thus improving photon emission from the vial.

As more water is added, the size of the dispersed water particles increases and an increase in self-absorption reduces the number of tritium beta particles escaping into the organic phase. The counting efficiency is therefore reduced. In contrast, the carbon-14 sample can be counted at constant efficiency in the "solution" phase and at near-constant efficiency in the emulsion phase because of the more energetic nature of the carbon-14 beta particles.

The partitioning of solutes between the organic and aqueous phases highlights a critical point in emulsion counting, which is that of efficiency determination. For emulsions, the internal standard should have the same partition coefficient for the emulsion system as that of the material being counted and, for tritium, this is normally tritiated water. In the case of carbon-14, it should be an aqueous standard solution of the material being counted. In any case, the internal standard must not be labeled toluene or hexadecane, which are insoluble in the aqueous phase. For a large number of samples, the sample channels ratio method for efficiency determination is convenient and useful if carefully done. However, the external standard cannot be used to monitor counting efficiency because the energy of the Compton electrons produced by the gamma source within the organic phase are essentially not attenuated by the dispersed water particles.

It has been observed that aqueous digests and extracts of tissues containing radioactivity incorporated into emulsion systems can sometimes be counted at equivalent (40) or higher efficiencies (41–43) than homogeneous samples. This interesting phenomenon can be ascribed to the probability that in emulsion counting the quenching substances remain primarily in the aqueous phase and do not interfere with the energy transferring process occurring in the organic phase. In a homogeneous solution, all components of the system are in such intimate contact that the effects of any quenching agent present are expressed.

Triton X-100 samples are notorious for their inconsistency when applied to liquid scintillation counting (36, 38, 43). In some instances, they contain impurities that luminesce, causing high and inconsistent background counts, and other batches have been known not to emulsify very well. The commercial preparations assure a dependable measure of performance and are formulated with surfactants similar to but not necessarily Triton X-100. There have been numerous reports on the use of commercial emulsifying agents (43–47). An inexpensive emulsion with high solubilizing power has been described by Anderson and McClure (48). It is formulated with 25% Triton X-114 in xylene containing 0.3%, (w/v), PPO as scintillator. If desired, a secondary scintillator can be added at a concentration of 0.2 g/liter. This system is capable of holding 2 ml of water in 5 ml of the Triton–xylene cocktail. The authors have found this system especially useful for acrylamide gels in which *N,N'*-diallyltartardiamide (DATDA) is substituted for *N,N'*-methyl bisacrylamide (BIS) as the cross-linking agent. Also, various solubilized tissues have been dissolved with this cocktail and counted at good efficiencies for both carbon-14 and tritium. It seems that the scintillator concentration of this cocktail is too low and that 6 g of PPO, w/v, may improve its performance. A similar cocktail containing naphthalene has also been suggested by Greene (49). Its formulation is 1 part Triton X-114 and 2 parts solution containing 100 g naphthalene, 10 g PPO, and 0.8 g POPOP per liter of xylene.

In summary, for successful emulsion counting, the following should be observed.

1. The purity of the surfactant is important. Different batches of commercial surfactants are not uniform and may contain luminescent impurities as well as quenchers.
2. All samples should be the same volume and have the same composition; i.e., the sample volume should be kept constant.
3. All samples should be prepared in the same way, e.g., heated to 40°C and allowed to cool without further agitation.
4. All samples should be equilibrated a minimum of 2 hours in the counter to stabilize the gel phase and temperature before counting. The counting properties of emulsions are sensitive to heat and constant temperature is required for reproducible counting. Dark adaptation also reduces any phosphorescence or chemiluminescence.
5. Either the sample channels ratio or the internal standard method should be used to determine counting efficiency.
6. External standardization cannot be used in emulsion counting.

MEASUREMENT OF $^{14}CO_2$

As all organic material can be oxidized to carbon dioxide, it is not surprising to find that much effort has been spent in developing systems suitable for $^{14}CO_2$ determinations. Indeed, measurement of $^{14}CO_2$ is now regarded as a routine procedure. The method chosen is primarily determined by the amount of gas involved. For small volumes, the $^{14}CO_2$ liberated in a closed system, such as a Warburg or combustion flask, is absorbed in a trapping agent and counted in a toluene cocktail. For large volumes of $^{14}CO_2$ the gas is often precipitated as $Ba^{14}CO_3$ and counted in suspension.

Hyamine hydroxide is the first agent to be described that forms a carbamate soluble in toluene. Although it is still widely used, better agents are now available. When present in excess, absorption of $^{14}CO_2$ is quantitative but the use of large quantities of Hyamine is inadvisable as it causes substantial quenching. Its high viscosity renders the quantitative transfer of Hyamine carbamate from the center well of a Warburg flask difficult but this problem can be overcome either by using a special flask in which the entire center well is removed and added directly to the counting vial or by using filter strips moistened with the trapping agent.

The most commonly used trapping agents are phenylethylamine and ethanolamine. Both compounds have greater capacity, are less quenching, and are more economical than Hyamine. Phenylethylamine is volatile and absorption of $^{14}CO_2$ is carried out at low temperatures. There have been doubts about the stability of the carbamate in toluene or dioxane counting solutions but, provided phenylethylamine is redistilled before use, it appears to be satisfactory. The carbamate of ethanolamine is soluble in ethyleneglycol monomethyl ether and can then be added to a toluene–PPO counting solution.

Other agents that can be used for trapping $^{14}CO_2$ are NCS, Soluene, and Protosol.

COMBUSTION TECHNIQUES

For some samples there are marked disadvantages to using direct solubilizing and suspension techniques. Primarily these center around the severity and type of quenching encountered. In addition, different types of samples of the same isotope may require different preparative procedures and utilize different counting solutions. The characteristics of the sample may also change on standing. One solution to these problems is to oxidize all samples of a particular isotope to a common material, such as carbon dioxide and water, and treat all the different samples in the same way. Combustion techniques therefore have wide application in reducing samples of diverse origins to a common denominator, and large or small samples can be counted with high efficiency in a reproducible manner. Internal standardization becomes unnecessary, the same instrument settings can be maintained throughout, and the samples are colorless and stable. However, the combustion techniques require special apparatus and some skill in handling. The sample is oxidized and the preparation of samples for counting is generally more time consuming than with other methods. Because the same apparatus is used repeatedly contamination is a potential hazard. Although accidents are rare, they are usually spectacular when they occur, and a safety shield between operator and apparatus is advisable.

Essentially there are two stages in the procedure: the oxidation itself and the absorption or removal of the end product. In the majority of cases the latter problem is usually one of counting carbon dioxide or water.

Oxygen Flask Methods

Most of the oxygen flask methods currently used are modifications of the Schoniger technique, in which the sample is placed in a closed flask filled with oxygen and ignited. Variations in the procedure center on the method

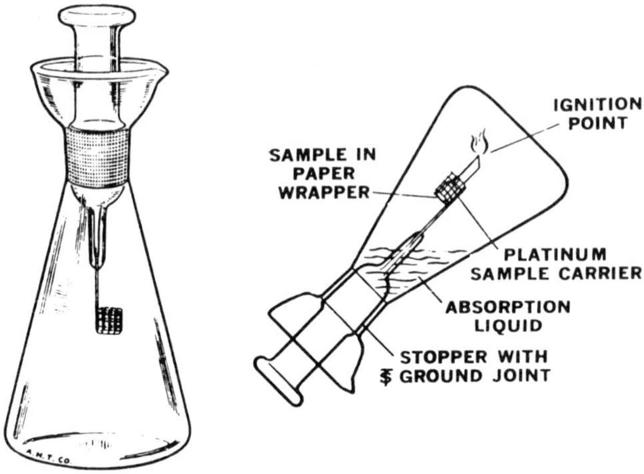

Fig. 35 A typical oxygen flask combustion apparatus.

of ignition and absorption of the combustion products. A typical oxygen flask is shown in Fig. 35.

The sample is dried in a cellophane bag or on filter paper and placed in a platinum or nichrome basket. After the basket is inserted, the flask (500 ml to 2 liters) is flushed with oxygen and closed. Ignition of the sample is then achieved by means of an electric spark passing through the basket or by focusing an infrared light source on the sample. The use of a filament or spark requires a more complicated flask head. As with most oxidation techniques, the sample must be dry; otherwise, difficulty is encountered in igniting the sample. The trapping agent is usually added after combustion is complete. Alternatively, a combustion flask with an extended tube at the bottom can be used. The trapping agent is added to the tube before combustion and immersed in a Dewar flask containing dry ice–acetone. The sample is then ignited and after oxidation is complete, the flask is removed from the Dewar flask and shaken for 20 minutes until the gas or water has been absorbed.

Oxygen Train Methods

This procedure is analogous to the classical method for carbon and hydrogen analysis. The dry sample is placed in a platinum basket with a catalyst and during heating oxygen is passed over the sample. The carbon dioxide is collected by bubbling the gas through a trapping agent and water is collected by freezing. The method is slow and contamination is a problem.

The furnace, however, can be designed to accommodate up to 2 g of material.

A commercial version of this apparatus is available and is called the Harvey Oxidizer; it is also sold by Beckman as the Beckman Biological Material Oxidizer (BMO). A liquid or solid sample is placed in a sample holder, which is introduced manually into the furnace. At the press of a button, the automated combustion cycle of 4 minutes is initiated. The sample is combusted in a hot, oxygen-enriched chamber maintained at 538°C, and the resulting gases are passed over the catalyst, copper oxide, maintained at 343°C to ensure complete oxidation of organic matter to carbon dioxide and water. It is claimed that the catalyst adsorbs all oxidation products of sulfur and halogens. The carbon dioxide is collected in an appropriate trapping agent and the water is collected either in a glass coldfinger or by bubbling the gas through a solution of Triton X-100 or Bio Solv-3 in a scintillator. The carbon dioxide and water traps must be changed and washed manually after each combustion. When 100 mg of labeled manitol is burned, recoveries of 97% for carbon-14 (specific activity >25 dpm/mg) and 90% for tritium (specific activity >200 dpm/mg) are claimed. This apparatus is essentially a manually operated unit.

Oxidation in a Sealed Tube

In this method the sample, weighing from 5 to 25 mg, is heated in a sealed, heavy glass tube along with copper oxide at 650°C for 1 hour. The tube is then placed in an apparatus in which oxidation is completed, the tube is broken, and the radioactive water vacuum is distilled into a dioxane solution. Relative to other combustion procedures the size of sample that can be oxidized is limited. Breaking tubes may be difficult and subsequent recovery of the combustion products variable. However, the method may be convenient for handling large numbers of small samples.

Wet Combustion Method

This procedure is based on the Van Slyke–Folch method for oxidizing biological samples in which a mixture of fuming sulfuric acid, periodate, and chromic acid is added to a flask containing the sample. After boiling for a couple of minutes the liberated $^{14}CO_2$ is passed through an alkaline solution. Although this method is of little value for tritium it can be used for the differential assay of carbon-14 and sulfur-35 in the same sample, as described below.

Oxygen Bomb Method

A Parr oxygen bomb can be used for the preparation of samples of low specific activities and up to 1 g of dry tissues can be combusted in this way. The apparatus is elaborate and the collection of the end products is time consuming. The method is rarely used.

One of the most powerful attributes of the scintillation counter is that it is possible to differentiate between two isotopes in the same sample provided their energy spectra are sufficiently different. Hence it is possible to count tritium in the presence of carbon-14. Nevertheless, complications often arise, particularly when quenching is severe and variable. Under these conditions, determining the amount of each isotope in the sample requires more stringent counting procedures. An advantage of combustion procedures is that the products can be separated and counted independently. By attaching a differential absorption train the sealed tube, oxygen train, and oxygen bomb methods can be adapted for separating tritium and carbon-14. The oxygen flask, which is probably the method of choice for singly labeled materials, usually requires more elaborate modification for dual-labeled samples.

A novel application of the oxidation techniques has been described for measuring carbon-14 and sulfur-35 in the same sample (50). For these two isotopes direct solubilization is of little value as their energy spectra are almost identical and cannot be differentiated by adjusting the pulse height analyzers. By oxidation of the sample in Pirie's reagent (a mixture of nitric and perchloric acids and magnesium nitrate), the isotopes can be separated chemically. Sulfur-containing material is oxidized to magnesium sulfate, which can then be solubilized in a counting solution containing glycerol, ethanol, dimethyl formamide, and toluene. The gaseous $^{14}CO_2$ can be trapped in the usual manner and measured without interference from sulfur-35.

AUTOMATED COMBUSTION APPARATUS

The great attraction of the combustion technique for sample preparation is the production of uniform samples from any type of biological specimen. However, the tedious and demanding nature of the procedure and its potential as a health hazard have inhibited its wide use. In recent years, three different semiautomatic combustion apparatuses have been developed and are commercially available. The most widely used is the Packard–Kaartinen apparatus (51).

Basically, Kaartinen has automated the oxygen flask system and has devised an apparatus that can be programmed to analyze a singly or doubly labeled sample containing either tritium and/or carbon-14. A flow diagram of the unit is shown in Fig. 36. For the analysis of a doubly labeled sample, the two collection vials for tritium and carbon-14 are pneumatically positioned, the tritium vial under the water condenser and the carbon-14 vial under the carbon dioxide absorption tower. The sample to be burned is carefully wrapped in a piece of filter paper and compacted in a press. The filter paper acts as fuel for the combustion and compaction allows the sample combustion to be better controlled. The compacted sample pellet is placed inside a platinum wire basket and the entire assembly is pneumatically positioned and sealed into a heated combustion chamber maintained at 180°C. To initiate the cycle, a button is pressed that automatically

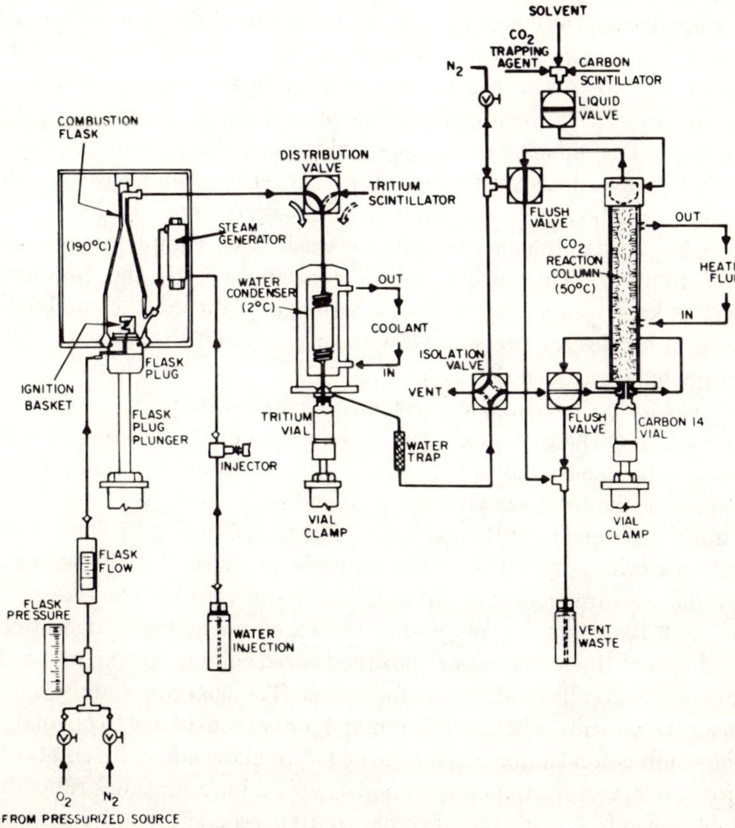

Fig. 36 Flow diagram of the Packard-Kaartinen combustion apparatus, model 305. (Courtesy Packard Instrument Company.)

charges the carbon dioxide absorption column with ethanolamine. Then the combustion is manually initiated by electrically heating the platinum sample holder and the speed of the burn is controlled manually by regulating the oxygen flow. The oxygen does not flow unless the oxygen valve knob is continuously depressed with the left hand. Manual control of the combustion sequence is a safety feature. When the oxygen flow is stopped at the completion of the burn, the "initiate cycle" button is pressed again and nitrogen is automatically flushed continuously through the entire system to sweep the remaining combustion products into the water condenser and the carbon dioxide absorption tower. Before the second cycle the combustion chamber is purged manually with three injections of approximately 70-μl aliquots of water that vaporize and act as carrier to ensure complete recovery of tritiated water formed during the burn. A predetermined amount of counting solution is delivered directly into the tritium vial. (In the original design, the counting solution is dispensed through the top of the condenser into the sample vial to help wash the condensed water out of the condenser and to reduce the carryover of radioactivity between sample burns. However, this arrangement has proved to be hazardous and unsatisfactory.) The radioactive carbon dioxide gas is absorbed by the ethanolamine that coats the absorption tower. The ethanolamine carbamate is sequentially washed into the sample vial with several methanol rinses followed by a final toluene–scintillator wash. One complete cycle takes about 2 minutes. The entire system contiguous with the combustion chamber is kept in a nitrogen atmosphere except during the combustion. The solvent bottles are pressurized with nitrogen and the solvents are dispensed pneumatically on demand.

Recovery for both tritium and carbon-14 is better than 95%. The tritium contamination of the carbon sample or the contamination of the tritium by carbon dioxide is less than 0.5%. The contamination percentage refers to the percent of the total activity of the offending isotope.

An improved version of the oxidizer is illustrated in Fig. 37. This second-generation oxidizer operates at atmospheric pressure and the water and carbon dioxide are collected at ambient temperature. The pressurized system with its associated plumbing, the requirement for refrigeration to trap and collect the water vapor, and the heated column to trap and collect the carbon dioxide have all been eliminated. The operator simply sets the dispensers to indicate whether tritium and/or carbon-14 are to be analyzed and the combustion timing clock is adjusted to allow adequate combustion time for the largest sample anticipated. Typical burning times range from about 30 seconds for an 0.5-g sample to 100 seconds for a 1.2-g sample, which is the largest recommended. To operate this system, two empty sample vials are positioned under the two condensers and the specimen to

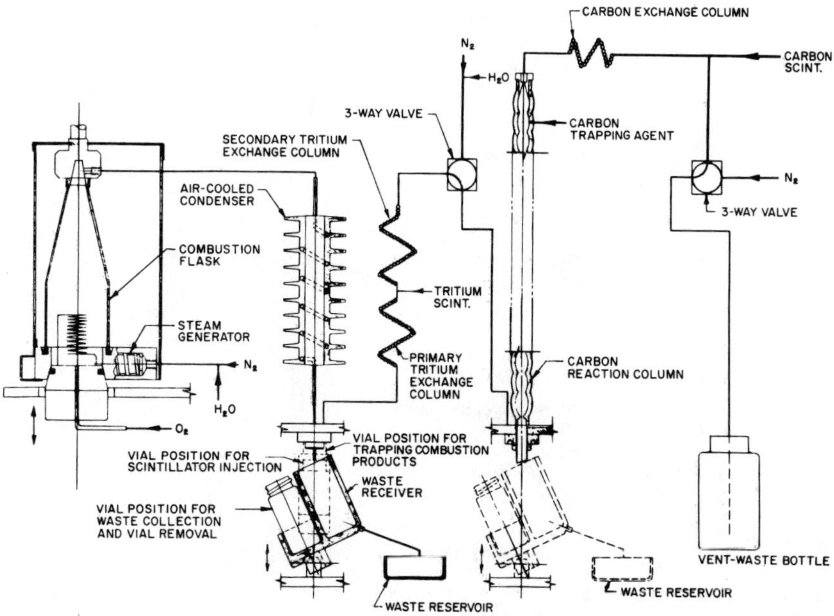

Fig. 37 Flow diagram of the Packard combustion apparatus, model 306. (Courtesy Packard Instrument Company.)

be burned is placed into the combustion basket. The specimen, as before, can be either pelletized in filter paper or put into a combustible sample holder. As soon as the sample is placed into the combustion basket, the "start" button is pressed. The combustion then proceeds automatically until completion. The sample is burned and the combustion flask purged with steam to remove all combustion products under a continuous flow of nitrogen. The water is condensed, trapped in the tritium exchange column, and quantitatively removed from the exchange column with a special scintillator solution by countercurrent flow and finally collected in the tritium vial. The carbon dioxide gas is insoluble in the scintillation solution and is swept into the carbon reaction column where it is trapped by a special trapping agent, by countercurrent flow. An exchange column ensures the quantitative recovery of carbon dioxide. The adsorbed carbon dioxide is then washed into the carbon-14 vial by the carbon dioxide scintillator solution.

The apparatus designed by Peterson and his group (52, 53) and marketed by Peterson is a semiautomated version of the oxygen train method for combusting organic samples. A flow diagram of the apparatus is shown in Fig. 38. The sample is placed in a 1-ml capsule of hardened gelatin. The capsule is dropped into the top of the apparatus into a "blind bore" stop-

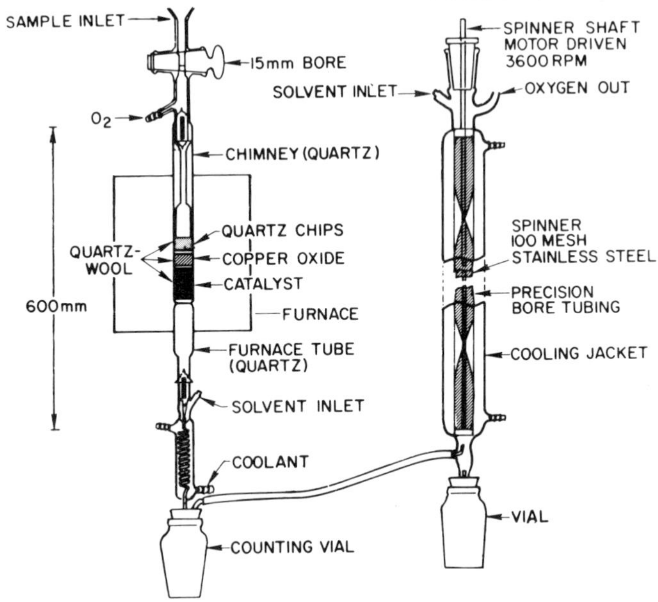

Fig. 38 The Peterson combustion apparatus with the tritium and carbon-14 absorption units.

cock. The operator presses the "solvent flow" button. At the same time, the stopcock is rotated 180° and the gelatin capsule drops by gravity through a chimney in which a stream of oxygen flows continuously and lands on a bed of quartz chips maintained at 600°C while the combustion chamber is simultaneously sealed from the atmosphere by the blind bore. The sample bursts into flames and the combustion products are carried in a stream of oxygen through a series of catalysts to complete the oxidation. The carbon dioxide and water leave the bottom of the furnace and enter a condenser tube. When only tritium is being analyzed, the condenser shown on the left side of Fig. 38 is used. Here, the combustion products encounter an incoming stream of cold scintillation solvent that condenses and dissolves the tritiated water. Twenty milliliters of the scintillation solvent at 3°C is pumped in automatically over a period of 2 minutes. The solvent is allowed to drain into a sample collection vial (a counting vial) for an additional minute. One sample may be processed every 3 minutes.

When carbon dioxide is to be collected, the carbon dioxide absorption apparatus shown on the right side of Fig. 38 is used. A Teflon delivery tube is connected to the tapered outlet tip of the combustion tube with a short piece of silicone rubber tubing to the bottom of the absorption tower. A spinner made from a stainless steel screen is rotated within the tower to maintain an even film of a carbon dioxide-absorbing liquid scintillation counting solution on the tower wall. The combustion gases entering the bottom of the tower flow against a stream of solvent entering at the top, assuring good contact between the gases and solvent. When carbon dioxide is to be collected, the solvent flow precedes the dropping of the sample. The sample is not dropped into the furnace until the flow of solvent has reached the bottom of the collection assembly, which takes approximately 2 ml of solvent.

The collection recovery for tritium is 96%, with a coefficient of variation of 2.9%; that for carbon dioxide is 97% with a coefficient of variation of 1.1%.

A flow diagram of the commercial version of the Peterson apparatus is shown in Fig. 39. The combustible sample is placed in a special polycarbonate capsule, which is loaded into a sample holder. For a doubly labeled sample, two sample vials are positioned under the proper ports. Then, the "start" button is pressed. This initiates the entire combustion and sample absorption program. The sample is automatically transported into the combustion furnace, maintained at 700°C, and is burned catalytically under oxygen. The water is scrubbed out of the combustion gases in a

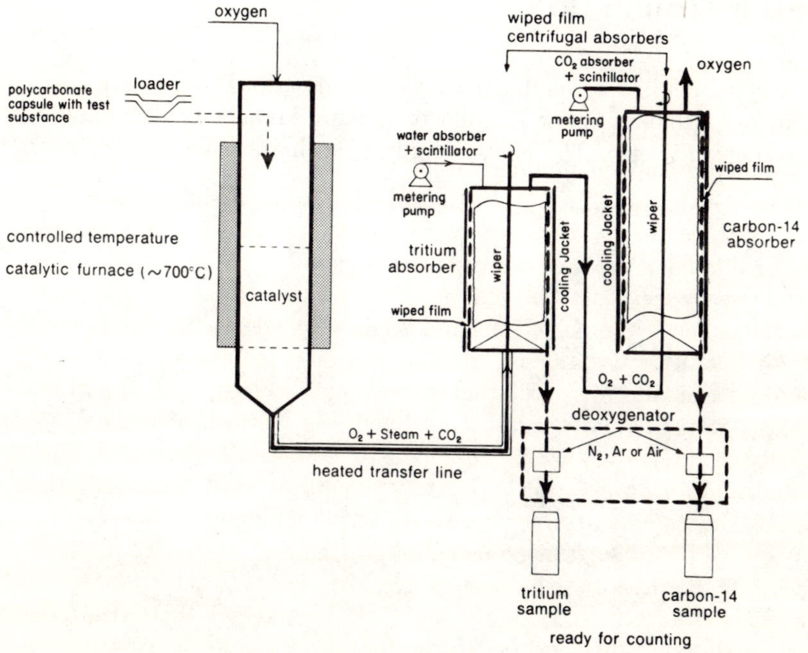

Fig. 39 Flow diagram of the Oxymat. (Courtesy Teledyne Intertechnique.)

water absorption tower with a counting solution containing a water absorber and the carbon dioxide is scrubbed out in a carbon dioxide absorption tower as in the Peterson apparatus. An added feature is a deoxygenator to reduce quenching in the final sample caused by oxygen. One complete cycle requires 3 minutes, allowing 160 samples to be combusted in an 8-hour day.

As a technique for sample preparation, combustion has proved to be a useful approach. However, the classical Schoniger technique, with its accompanying problems of safety, the tedious nature of the process, the limitations of sample size, and the small number of samples that can be analyzed in a single day because of the glassware required have all contributed to discourage its use. All of these problems can be eliminated by using one of these semiautomatic combustion apparatus. Samples can be burned continuously with complete safety with near-quantitative recoveries of radioactivity at an average throughput time of 3 minutes. The resultant samples are uniform, clear, and colorless solutions that can be accurately assayed in a liquid scintillation counter. Assuming no mechanical problems, the major obstacle to the wide use of these instruments is probably the relatively high capital cost.

SPARK COMBUSTION

A novel method of combustion has been described by Noakes (54) and utilizes continuous spark ignition of organic samples in disposable steel planchets (Fig. 40). The power supply is an inexpensive battery charger

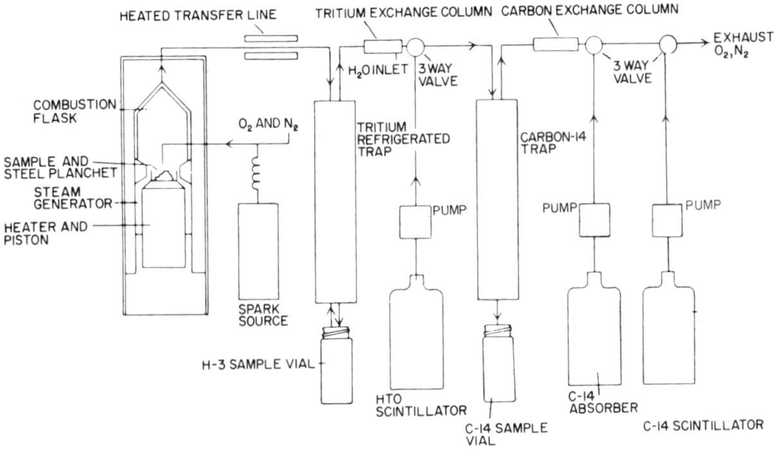

Fig. 40 Flow diagram of the spark combustion apparatus (54).

and the spark coil is an ignition coil used in automobiles. The sample is combusted in the following manner. The sample is placed into a steel planchet, which is then positioned on a heated piston that is also an electrical conductor and acts as the ground (earth) for the spark. The piston is secured in the combustion chamber and the ignition started. Oxygen is fed through a stainless steel tube located directly over the sample and the spark is carried by the oxygen tube, which doubles as the positive electrode. The spark ignites on the organic material and the oxygen flow maintains smooth combustion. Sparking is continued until the entire sample is oxidized. Figure 40 shows that the water and carbon dioxide trapping systems are similar to those previously described. Using tritiated and carbon-14 labeled thymidine, recoveries of 97.8 ± 1.5% for tritium and 98.5 ± 1% for carbon-14 are reported. Spillover and memory for tritium are 0.8 and 0.2%, respectively, and for carbon-14, 0.07 and 0.05%, respectively.

SUSPENSION COUNTING

The severe limitation imposed by the solubility of samples in toluene-based counting solutions has been overcome in many cases by the gel counting technique. The general procedure is to prepare a sample of uniform particle size and to suspend it in a solvent system containing a gelling agent, such as aluminum stearate, thixin, or Cab-O-Sil. Thixin is a castor oil derivative and Cab-O-Sil a finely divided silica. The latter is the most widely used. Maximum counting efficiencies are obtained using 3.5–4.0% Cab-O-Sil. Below this concentration the gel is unstable and at higher concentrations the silica particles interfere with the transmission of light, causing a subsequent reduction in counting efficiency.

For solid materials the method requires that the sample be finely powdered and that it be white or colorless to avoid light absorption. The sample should not dissolve in the scintillation fluid; otherwise, the dissolved material is counted at a different efficiency from the suspended material. The procedure is simple to perform, large amounts of sample can be counted, and solvent quenching by the sample is minimized because the material is not in solution.

Counting efficiency, however, may be reduced by the following.

a. Settling of the suspension causing a departure from 4π geometry
b. Opacity of the resulting medium, which increases light scattering
c. Self-absorption

The most interesting and problematical of these factors is that of self-absorption. For tritium the self-absorption losses may be so large that the technique is sometimes of little value. For samples that are nonuniform in particle size the extent of self-absorption may vary between samples. It is therefore necessary under these circumstances to sieve or grind the sample to ensure a constant particle size. As this procedure is time consuming it constitutes one of the major disadvantages of the method. Fortunately, in the case of barium carbonate particle size is not important. Self-absorption losses are still present but are constant between samples. The losses caused by self-absorption may be determined by counting in suspension a material of known specific activity and the losses may be estimated by the difference. The counting efficiency of that radiation which does interact with the scintillator may be determined by any of the methods used for homogeneous solution counting.

One of the widest applications of the suspension technique has been for the determination of barium carbonate and this is often used when large volumes of $^{14}CO_2$ are involved. The method has also been used for aqueous solutions and tissue homogenates, although better methods are now available. Cab-O-Sil can be used to reduce the adsorption of radioactivity onto the glass walls of a counting vial. If Cab-O-Sil is added first, the radioactivity is adsorbed onto the suspended particles in preference to the walls of the vial by virtue of its larger surface area. For some compounds the increase in counting efficiency is substantial. The suspension technique has also been used for measuring radioactivity in thin-layer chromatograms. Scrapings from the chromatogram are simply mixed with the gel. The method is unsuitable, however, if partial elution from the binding support takes place.

COUNTING ON A SUPPORT MEDIUM

An alternative method for dealing with samples insoluble in the counting solution is to deposit them on a support medium, such as paper strips, filter disks, glass fiber, or DEAE cellulose, before adding them to the vial. An obvious use of this is in chromatography, wherein the paper can be cut into strips and added to the counting solution. The procedure is simple and useful for materials insoluble in scintillator fluid. As with the suspension methods quenching is diminished as the agent is not in solution with the solvent. However, reproducibility can only be obtained if the sample is totally insoluble in the counting solution. Variable results are obtained if partial elution occurs. This can be determined by removing the support medium from the vial and counting the residual activity in the scintillator. Difficulties may also be encountered with colored samples and, unless the support medium is thin and translucent, differences in penetration into the medium may also cause variable results. For tritium, particularly, self-absorption may be a problem, as it is with the suspension techniques, and there is controversy over whether orientation of the supporting medium is important. For carbon-14 this is not usually a problem but if old instruments are used, reproducible orientation, such as placing the strip on the bottom of the vial, may be necessary for tritium samples. An advantage of this counting technique over most others is that if the sample is insoluble it can easily be recovered. The strips can be washed with toluene to remove the scintillator, dried, and then eluted with an appropriate solvent. Another approach of current interest is to use a support medium impregnated with a plastic scintillator, such as Naton 136 (55). The closer contact of the sample with the scintillator gives rise to a much higher counting efficiency for tritium.

CERENKOV COUNTING

When a solution contains a high concentration of an energic beta emitter, such as phosphorus-32, the solution emits a faint, bluish-white light, known as "Cerenkov radiation." Cerenkov light is generated when a charged particle travels through a medium faster than the speed of light through that medium. The emitted light is directional and cone shaped with a half-angle ϕ. Cosine ϕ is defined as follows (56).

$$\cos \phi = 1/\beta n$$

where

$$\beta = (\text{velocity of particle})/(\text{speed of light}) = v/c$$
$$n = \text{refractive index of medium}$$

The threshold electron energy, $E_{\min}$, for Cerenkov radiation can be computed from the relationship (22)

$$\beta n = 1$$

where E is related to β by the equation

$$\beta = \left[1 - \left(\frac{1}{E_{(\text{keV})}/511 + 1}\right)^2\right]^{1/2}$$

In water, which has a refractive index of 1.332, the minimum electron energy for Cerenkov radiation is 263 keV. If it is assumed that the medium is nondispersive, there is a continuous spectrum of radiation emitted, with the greatest number of photons being produced in the ultraviolet and extending into the visible but negligible in the infrared. Cerenkov radiation is distinct from either fluorescence or *bremsstrahlung* and is very rapid, the duration of the light pulse being less than 1 nsecond.

In a 1964 review article, Rapkin (57) suggested that phosphorus-32 might be best counted by measuring Cerenkov radiation. He cited the unpublished work of DeVolpi and Horrocks (1964) on the measurement of manganese-56 ($\beta^- = 2.81$, 1.04, and 0.65 MeV_{max}) in an aqueous sulfate solution by measuring Cerenkov radiation directly in a coincidence liquid scintillation counter without the addition of any scintillator. Counting efficiency using only a single photomultiplier tube was 90% and only 35% in coincidence.

Coincidence counters have their photomultiplier tubes at an angle of 180° (in line), which is a disadvantage for detecting the highly directional Cerenkov light. This is illustrated in Fig. 41, which shows the angular photon distribution of Cerenkov light generated by ^{170}Tm in water as determined by Ross (58, 59). As in normal liquid scintillation counting, the addition of a wavelength shifter has been found efficacious in Cerenkov counting. The shifter not only reemits the absorbed ultraviolet light in the visible region but also isotropically (in all directions), thus markedly enhancing its detection by a coincidence counter (Fig. 41).

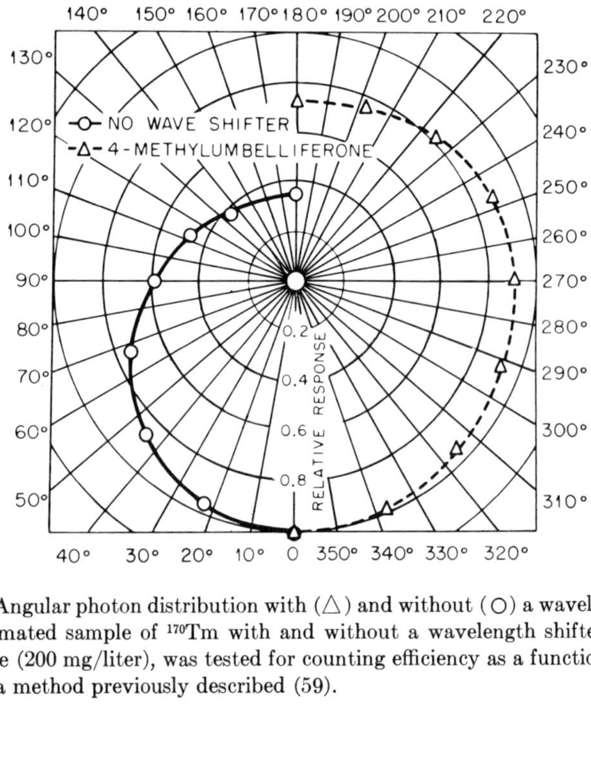

Fig. 41 Angular photon distribution with (△) and without (○) a wavelength shifter (56). A collimated sample of ^{170}Tm with and without a wavelength shifter, 4-methylumbelliferone (200 mg/liter), was tested for counting efficiency as a function of angular rotation by a method previously described (59).

Theoretically, a beta emitter must have an energy greater than 263 keV to be detected in water by Cerenkov counting. This eliminates tritium, carbon-14, and sulfur-35 from consideration for this technique. The minimum energy threshold, however, is a function of the refractive index of the medium. Therefore, Ross (22) has calculated that a medium with a refractive index greater than 1.559 may allow the detection of carbon-14. With α-bromonaphthalene, $n = 1.6582$, the detection efficiency is 0.1%. When the sample has actually been tested, the detection efficiency is a surprising 12.3%. Ross has suggested that this unexpected result arises because the refractive index of α-bromonaphthalene may be greater than originally thought, because this system is operating in the region of anomalous refractive dispersion. If the counting efficiency for carbon-14 can be further increased, then Cerenkov counting may become important for this isotope.

Cerenkov counting is currently useful for energetic beta emitters. Parker and Elrick (56) have experimentally determined the Cerenkov counting

TABLE 7

Experimentally Determined Cerenkov Counting Efficiencies

		Counting efficiency (% of disintegration)[a]		
		S11 Photocathode		RCA 4501-V3 photo-cathode,
Nuclide	E_{max} (MeV)	Aqueous sample	Wavelength shifter	aqueous sample
^{204}Tl	0.77 (98%)	1.3	2.6	—
^{137}Cs	0.51 (92%), 1.17 (8%)	2.1	—	—
^{36}Cl	0.71 (98.3%)	2.3	4.7	5.3
^{198}Au	0.96 (99%)	5.4	—	—
^{47}Ca	0.66 (83%), 1.94 (17%)	7.5	14.8	—
^{40}K	1.32 (89%)	14	31	34.3
^{24}Na	1.39 (100%)	18	40	—
^{86}Rb	0.68 (8.5%), 1.77 (91.5%)	23	46	—
^{32}P	1.71 (100%)	25	50	43.3
^{144}Ce–^{144}Pr	2.98 (97.7%)	54	75	59.2
^{42}K	2.0 (19%), 3.6 (82%)	60	85	76.3
^{106}Ru–^{106}Rh	2.0 (3%), 2.4 (12%) 3.1 (12%), 3.6 (70%)	62	85	75.3

[a] Comparisons are given with and without the water-soluble sodium–potassium salt of 2-naphthylamine-6,8-disulfonic acid (100 mg/liter), using now obsolete photomultipliers. The values given for more modern phototubes are in the absence of wavelength shifter; if shifter is added, even higher efficiencies may be expected. (56).

efficiencies of a number of beta-emitting radioisotopes in a 10-ml solution (Table 7). The wavelength shifter used in Table 7 is the water-soluble salt of 2-naphthylamine-6,8-disulfonic acid (100 mg/liter). Should the wavelength shifter be used with the RCA tube, higher efficiencies than those given can be observed.

As a means of analysis, Cerenkov counting offers many advantages. It is a simple technique because the sample can be counted directly without the addition of any scintillator. The sample size is limited only by the capacity and geometry of the sample counting chamber and is not subject to chemical quenching. The sample is not contaminated by extraneous chemicals and the solvent can be almost any colorless liquid.

The disadvantages of the method are that the counting efficiencies are relatively low and it is sensitive to color quenching. However, the poor counting efficiency can be increased by adding a wavelength shifter and color quenching can, in most instances, be removed by chemical treatment.

The most useful application in the biological sciences is the measurement of phosphorus-32, which can be counted directly at greater than 40% efficiency in a counter equipped with bialkali photomultiplier tubes (Table 7). Higher counting efficiencies can be realized by adding a wavelength shifter and by using plastic vials. The latter help scatter the light more efficiently and also reduce the background. The advantages of Cerenkov counting of phosphorus-32 in photophosphorylation research have been discussed by Ould *et al.* (60).

REFERENCES

1. Kallmann, H., and Furst, M., *In* "Liquid Scintillation Counting" (C. G. Bell and F. N. Hayes, eds.), p. 3. Pergamon, Oxford, 1958.
2. Furst, M., Kallmann, H. and Brown, F. H. *Nucleonics* **13,** 58 (1955).
3. Hayes, F. N., Ott, D. G., Kerr, Y. N., and Rogers, B. S. *Nucleonics* **13,** 38 (1955).
4. Ott, D. G., Hayes, F. N., Hansbury, E., and Kerr, V. N. *J. Amer. Chem. Soc.* **79,** 5448 (1957).
5. Ott, D. G., Kerr, V. N., Hayes, F. N., and Hansbury, E. *J. Org. Chem.* **25,** 872 (1960).
6. Nuclear Chicago Radiochemical Bulletin No. 34 (1967).
7. Bush, E. T., and Hansen, D. L. *In* "Radioisotope Sample Measurement Techniques in Medicine and Biology," pp. 395–408. IAEA, Vienna, 1965.
8. Swank, R. K. *In* "Liquid Scintillation Counting" (C. G. Bell and F. N. Hayes, eds.), p. 23. Pergamon, Oxford, 1958.
9. Kobayashi, Y. *Anal. Chem.* **38,** 1240 (1968).
10. Litle, R. L., and Neary, M. P. *In* "Organic Scintillators and Liquid Scintillation Counting" (D. L. Horrocks and C. Peng, eds.), p. 431. Academic Press, New York, 1971.

11. Hayes, F. N., Rogers, B. S., Sanders, P., Schuch, R. L., and Williams, D. L. Los Alamos Sci. Lab. Reports LA-1639 (1953).
12. Hayes, F. N., Rogers, B. S., and Sanders, P. C. *Nucleonics* **13,** 46 (1955).
13. Kerr, V. N., Hayes, F. N., and Ott, D. G. *Int. J. Appl. Radiat. Isotop.* **1,** 284 (1957).
14. Davidson, J. D., and Feigelson, P. *Int. J. Appl. Radiat. Isotop.* **2,** 1 (1957).
15. Furst, M., and Kallmann, H. *Phys. Rev.* **85,** 816 (1952).
16. Bray, G. A. *Anal. Biochem.* **1,** 279 (1960).
17. Benson, R. H., and Maute, R. L. *Anal. Chem.* **34,** 1122 (1962).
18. Gordon, B. E., and Curtis, R. M. *Anal. Chem.* **40,** 1486–1493 (1968).
19. Blanchard, F. A., and Takahashi, I. T. *Anal. Chem.* **33,** 975 (1961).
20. Petroff, C. P., Nair, P. P., and Turner, D. A. *Int. J. Appl. Radiat. Isotop.* **15,** 491 (1964).
21. Eichholz, G. G., Nagel, A. E., and Hughes, R. B. *Anal. Chem.* **37,** 863 (1965).
22. Ross, H. H. *In* "The Current Status of Liquid Scintillation Counting" (E. D. Bransome, Jr., ed.), p. 123. Grune & Stratton, New York, 1970.
23. Boyce, I. S., and Cameron, J. F. *In* "Tritium in the Physical and Biological Sciences," Vol. 1, p. 231. IAEA, Vienna, 1962.
24. Hayes, F. N. *In* "Liquid Scintillation Counting" (C. G. Bell, Jr. and F. N. Hayes, eds.), p. 83. Pergamon, Oxford, 1958.
25. Birks, J. B. "The Theory and Practice of Scintillation Counting." Pergamon, Oxford, 1964.
26. Winkelman, J., and Slater, G. *Anal. Biochem.* **20,** 365 (1967).
27. Kalbhen, D. A. *Int. J. Appl. Radiat. Isotop.* **19,** 49 (1968).
28. Gower, M. F., and Bransome, E. D., Jr. *In* "The Current Status of Liquid Scintillation Counting" (E. D. Bransome, Jr., ed.), p. 263. Grune & Stratton, New York, 1970.
29. Dobbs, H. E. *Int. J. Appl. Radiat. Isotopes,* **19,** 155 (1968).
30. Mahin, D. T., and Lofberg, R. T. *In* "The Current Status of Liquid Scintillation Counting" (E. D. Bransome, Jr., ed.), p. 212. Grune & Stratton, New York, 1970.
31. White, D. R. *Int. J. Appl. Radiat. Isotop.* **19,** 49 (1968).
32. McEvoy, A. F., Dyson, S. R., and Harris, W. G. *Int. J. Appl. Radiat. Isotop.* **23,** 338 (1972).
33. Painter, K., and Gezing, M. J. *Int. J. Appl. Radiat. Isotop.* **24,** 361 (1973).
34. Turner, J. C. *Int. J. Appl. Radiat. Isotop.* **18,** 557 (1968).
35. Turner, J. C. *Int. J. Appl. Radiat. Isotop.* **20,** 499 (1969).
36. Patterson, M. S., and Greene, R. C. *Anal. Chem.* **37,** 854 (1965).
37. Benson, R. H. *Anal. Chem.* **38,** 1353 (1966).
38. van der Laarse, J. D. *Int. J. Appl. Radiat. Isotop.* **18,** 485 (1967).
39. Williams, P. H., and Florkowski, T. *In* "Radioactive Dating and Methods of Low Level Counting," p. 703. IAEA, Vienna, 1967.
40. Greene, R. C., Patterson, M. J., and Estes, A. H. *Anal. Chem.* **40,** 2035 (1968).
41. Madsen, N. P. *Anal. Biochem.* **29,** 542 (1969).
42. Fox, B. W. *Int. J. Appl. Radiat. Isotop.* **19,** 717 (1968).
43. Chapman, D. I., and Mancroft, J. *Int. J. Appl. Radiat. Isotop.* **22,** 371 (1971).
44. Wiebe, L. I., Stevens, A., Norijaim, A. A., and Ediss, C. *Int. J. Appl. Radiat. Isotop.* **22,** 663 (1971).
45. Francois, B., and Limandas, M. *Int. J. Appl. Radiat. Isotop.* **22,** 632 (1971).
46. Bohne, F. *Int. J. Appl. Radiat. Isotop.* **22,** 384 (1971).
47. Friedman, M. A., Millar, G., McEvoy, A., and Epstein, S. S. *Anal. Chem.* **43,** 780 (1971).

48. Anderson, L. E., and McClure, W. O. *Anal. Biochem.* **51,** 173 (1973).
49. Greene, R. C. *In* "The Current Status of Liquid Scintillation Counting" (E. D. Bransome, Jr., ed.), p. 189.Grune & Stratton, New York, 1970.
50. Jeffay, H., and Alvarez, J. *Anal. Biochem.* **2,** 506 (1961).
51. Kaartinen, N. Packard Technical Bulletin No. 18. Packard Instrument Co., Inc., Downers Grove, Illinois, 1969.
52. Peterson, J. I., Wagner, F., Siegel, S., and Nixon, W. *Anal. Biochem.* **31,** 189 (1969).
53. Peterson, J. I. *Anal. Chem.* **31,** 204 (1969).
54. Naokes, J. E. *In* "Advances in Liquid Scintillation Counting" (P. E. Stanley and B. Scoggins, eds.), p. 53. Academic Press, New York, 1974.
55. Nakshbandi, M. M., *Int. J. Appl. Radiat. Isotop.* **16,** 157 (1965).
56. Parker, R. P., and Elrick, R. H. *In* "The Current Status of Liquid Scintillation Counting," (E. D. Bransome, Jr., ed.), p. 110. Grune & Stratton, New York, 1970.
57. Rapkin, E. *Int. J. Appl. Radiat. Isotop.* **15,** 69 (1964).
58. Ross, H. H., *Anal. Chem.* **41,** 1260 (1969).
59. Ross, H. H. *In* "Organic Scintillators and Liquid Scintillation Counting" (D. L. Horrocks and C. Peng, eds.), p. 757. Academic Press, New York, 1971.
60. Ould, J. M., Cather, R., and Winget, G. D. *Anal. Biochem.* **50,** 540 (1972).

4

Testing and Setting up a New Counter

GENERAL CONSIDERATIONS

It is good practice to install a liquid scintillation counter away from all high-energy laboratories that have such equipment as neutron generators and X-ray machines and from laboratories that handle large quantities of gamma emitters, such as iodine-131. Even though all liquid scintillation counters are well shielded for cosmic radiation, local sources of radiation can cause extraneous, unwanted counts to be added to the sample count rate. If the new counter is a temperature-controlled model, careful attention should be paid to the lid of the sample compartment to see that it fits properly. The temperature of these units should be set between 5° and 10°C to prevent aqueous samples from freezing and also to minimize the accumulation of ice on the walls of the sample chamber, especially during humid weather. Ice can build up on the walls to a point where it impedes the transport of samples, causing either the sample belt to stretch and break or the sample conveyor to stop, which in turn may cause a fuse to fail or the sample belt drive motor to burn. In a modern counter, the primary reason for controlled temperature is to assure constant performance because all photomultiplier tubes have a temperature coefficient (i.e., change in operating efficiency with temperature). Bialkali photomultiplier tubes become very inefficient at temperatures below −10°C so instruments equipped with these tubes are usually operated above 0°C.

In testing a new unit, the primary considerations are the accuracy of the discriminators, the reproducibility of the gain or attenuator controls of counters with linear amplifiers, and the stability of the instrument as a whole. The stability of the counter, however, cannot be tested until the unit has been allowed to equilibrate at least 3 days in its new environment. During this time, the heavy shielding around the photomultiplier tubes comes to thermal equilibrium, the photomultiplier tubes dark adapt, and the electronics "bake in." The performance of the counter generally improves as it ages, particularly with respect to background. This period of equilibration can be utilized to check the mechanical functions of the counter, such as the sample changer, the group selection mechanism, the anti-sample-jam device, the printer or teletype, and such electronic paraphenalia as the background subtract unit, the low-count reject function, the repeat count function, the automatic external standard unit, and the data computer. Special attention must be paid to those units which are equipped with a teletype and tape punch. The punched tape prepared by the teletype must be error-free if it is to be of any value for data processing by computer.

With multichannel liquid scintillation counters, gain controls and discriminators should be equivalent in all channels; that is, two channels set at identical gains and discriminator settings should give similar results. This is required for accurate and reproducible pulse height analysis. In a three-channel unit, the third channel can be a spare channel in the event another fails but, more important, it can be used to monitor counting efficiency using the sample channels ratio technique.

As a convenience feature, most manufacturers now offer preset discriminators that can be activated by pushbutton controls, by a selector switch, or by simply replacing the variable discriminator plug-in module with a preset discriminator module. The gain controls of the linear amplifier units and discriminators of all types of new counters can be tested while the instrument is in the process of environmental equilibration.

TESTING COUNTERS BASED ON LINEAR PULSE HEIGHT ANALYSIS

Gain or Attenuator Tracking Test

Use a sealed tritium reference sample in this test. Set the lower disscriminator at 2 and the upper discriminator at 100 in all channels (100 is taken as the maximum value). Set all gain controls at minimum value or attenuators at their maximum value. Take a 1-minute count. Now, increase the gain by 10% or reduce the attenuator by one increment. Take a 1-minute count, and repeat this procedure until 100% gain (maximum) or the minimum attenuation is achieved. All channels should count within 1% of each other.

Balance Point Determination

When the results of the gain tracking test are examined, it is seen that the counts gradually increase as the gain is increased until a maximum is reached. With a further increase in gain, the counts plateau and then decline. The gain at which the plateau occurs is the balance point for the discriminator settings used and represents the most stable point for counting tritium. The exact position of the balance point can be found by manipulating the fine gain or attenuator control in the appropriate direction. The discriminator settings of 2–100 are sometimes referred to as a 50:1 window and represent one with a relatively high dynamic range. In the case of tritium, such a wide window usually gives the highest figure of merit. When the window width is changed, the balance point also changes.

Discriminator Tracking Test

Use the tritium sample in this test. Set all the gain controls at the balance point found in the gain tracking test. In this test, the discriminators are varied similarly in all channels to show that they are equivalent. One-minute counts are sufficient if the sample contains over 100,000 dpm. The discriminators are varied according to the schedule given in Table 8 and all channels should count within 1% of each other. If the counts are not tracking within 1% adjust the lower discriminator. The linearity of

TABLE 8

Discriminator Tracking Test Schedule for Linear Units

Trial	Lower discriminator	Upper discriminator
1	2	20
2	20	40
3	40	60
4	60	80
5	80	100

potentometers used for the discriminators are guaranteed only between 5 and 95% of the entire range (5–95 divisions).

Balance Point for Carbon-14

If the gain and discriminator tracking tests are satisfactory for tritium, it can be assumed that they are so for carbon-14. However, the best figure of merit for carbon-14 is usually obtained with a narrower window than that required for tritium. Therefore, the balance point for carbon-14 should be obtained with about a 10:1 window, i.e., with the lower discriminator at 10 and the upper discriminator at 100. The gain setting should again be set at its minimum. The successive gain increases should be in 1% increments up to 10%. The balance point occurs at about 5% gain and the counting efficiency here should exceed 90%. The exact position of the balance point is obtained by adjusting the fine gain control. If the instrument has attenuators instead of gain controls, the attenuators should be at their maximum and should be reduced stepwise after each 1-minute count. When the balance point is indicated by the high count rate, its exact position can be found by adjusting the fine attenuator control in the appropriate direction.

Isotope Separation Test

In a double-isotope analysis, the lower energy isotope (tritium) can be separated from the higher energy isotope (carbon-14), but never the reverse. This can be shown by establishing how much of the carbon-14 spectrum lies above the tritium spectrum. This is accomplished by first manipulating the gain (or attenuator) so that the carbon-14 spectrum occupies the full window (from 0 to 100). This is done as follows: First, an unquenched carbon-14 standard is loaded and held in the counting chamber (by using either the manual sample control or the repeat count mode). The lower discriminator is set at 100 and the pulse height analyzer mode switch is set at the "lower to infinity" position. The counter is started and the gain control is increased, starting from zero, until counts just begin to appear in the window. At this gain setting, the carbon-14 spectrum occupies the full window. The upper discriminator is set at 100 and the pulse height analyzer mode switch is set back to the "lower to upper" position. An unquenched tritium sample is now loaded and held in the counting chamber. The upper and lower windows are set at 100. The counter is started and the lower discriminator is reduced from 100 until the tritium counts just appear in the window. The lower window is adjusted so that less than 0.01% of the tritium appears in the window. This point is difficult to determine because the count rate is near the background level. At this point, the carbon-14 sample is again counted and the computed efficiency should be between 60 and 65%. If these performance figures are met, the instrument has demonstrated acceptable behavior with regard to isotope separation.

The tritium channel can be set up in the following manner to obtain the highest tritium efficiency with the lowest concomitant carbon-14 efficiency. Adjust the gain in the tritium channel to its maximum (or the attenuator at its lowest position). Set both the upper and lower discriminators at 2. Load the unquenched carbon-14 sample and start counting. Raise the upper discriminator until the carbon-14 efficiency reaches 6 ± 0.5%. Next, load the unquenched tritium sample and determine its counting efficiency. It should be 35 ± 5%. The relationship between carbon-14 and tritium counting efficiencies, determined experimentally under conditions of maximum gain and using air-quenched samples of labeled toluene containing 0.8% butyl-PBD, is shown in Fig. 42. The curve has been obtained in the following manner: The gain of the counter is adjusted to maximum gain (or least attenuation). The lower discriminator is set at 2 divisions and the upper at 20 divisions. Both the carbon-14 and tritium samples are counted. The upper discriminator is then raised in 5-division increments and the two counts repeated at each new upper discriminator setting until the full window width is reached. This type of plot can be used to determine the best counting efficiency ratio possible between tritium and carbon-14, which is the steepest tangent drawn from the base of the curve. In this case, this is at about 4% counting efficiency for carbon-14 and at about 26% for

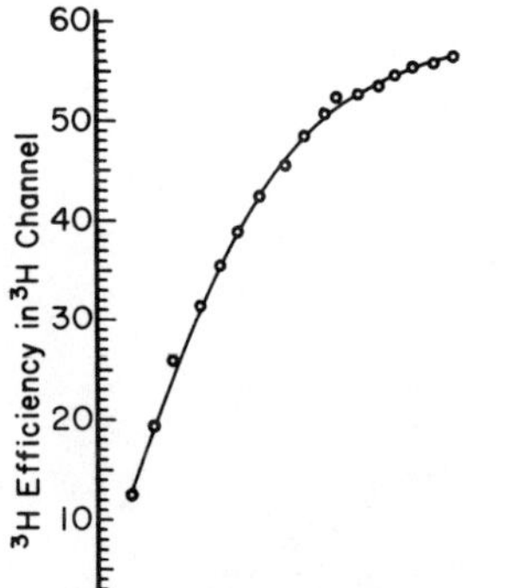

Fig. 42 Relationship between the counting efficiencies of carbon-14 and tritium with increasing tritium efficiencies in the tritium counting channel. Air-quenched samples of [^{3}H]- and [^{14}C]toluene containing 0.8% butyl-PBD in 10 ml of toluene were analyzed in a counter with linear amplification.

tritium, a ratio of 6.5. However, a more practical choice may be a compromise to facilitate tritium counting by taking about 6% carbon-14, which gives about 34% counting efficiency for tritium and a $^{3}H/^{14}C$ efficiency ratio of 5.7.

Figure 43 shows experimentally determined spectra of carbon-14 and tritium using air-quenched samples in toluene containing 4 g of PPO per liter. The spectra have been plotted after first adjusting the gain so that the carbon-14 spectra occupies the full window width of a 100 divisions, as described above. The entire tritium spectrum lies below 17 divisions and more than 50% of the carbon-14 spectrum lies above 17 divisions. This means that above 17 divisions, carbon-14 can be counted without any contribution from tritium in a doubly labeled sample. This method of analysis is known as the "isotope exclusion method." The advantages of this procedure are that the carbon-14 content of a doubly labeled sample can be determined very accurately, the calculations for the contribution for each isotope are simplified, and the error of the carbon-14 correction in the tritium channel is minimized. Close examination of the spectra shown in Fig. 43 shows, however, that between 10 and 17 divisions there is only a small amount of tritium. Therefore, by allowing a small amount of tritium to appear in the carbon-14 spectrum, the carbon-14 counting efficiency can be significantly increased. With no tritium in the carbon-14 spectrum, the carbon-14 counting efficiency is about 51%. By including 0.1% of the

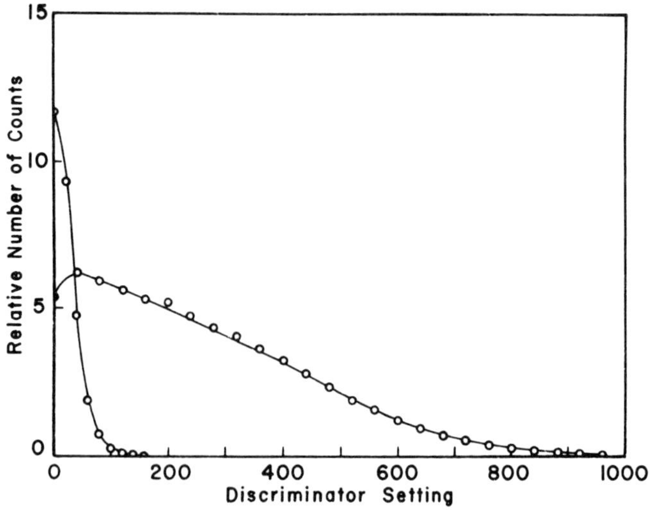

Fig. 43 The beta spectra of carbon-14 and tritium determined experimentally in a counter with linear amplification. Ten milliliters of an air-quenched sample of [^{3}H]- and [^{14}C]toluene containing 4 g of PPO per liter of toluene were used.

TABLE 9

Effect of Increasing Tritium Spillover into the ^{14}C Counting Channel on ^{14}C Counting Efficiency[a]

^{3}H Counting efficiency (%)	^{14}C Counting efficiency (%)
0	59.8
0.001	60.8
0.005	64.3
0.01	65.9
0.05	69.3
0.1	70.8
0.5	74.3
1.0	75.8

[a] Data points from an air-quenched sample of [^{3}H]- or [^{14}C]-toluene containing 0.8% butyl-PBD, using a counter with linear amplification and quartz-faced bialkali photomultiplier tubes.

tritium activity, the carbon-14 can be increased about 15%. With 1% tritium, the counting efficiency for carbon-14 is about 72%. As a comparison, a similar determination has been made in a counter using bialkali photomultiplier tubes and the data are presented in Table 9. The excellent performance of these tubes is demonstrated by the high carbon-14 efficiency of near 60% in the absence of tritium in an air-quenched toluene sample containing 0.8% butyl-PBD.

Engberg Plots

The test for isotope separation has been described for the isotope pair carbon-14 and tritium because this is the most common pair encountered in biological research. An alternate way to graphically characterize the isotope separation performance of a counter is to make an Engberg plot, named after its originator. The Engberg plot is constructed by plotting the log of the carbon-14 efficiency against the log of the tritium efficiency in a fixed window (usually, a wide window of at least 25:1) and then counting alternately carbon-14 and tritium at gradually increasing gain from zero to maximum. An Engberg plot for air-quenched samples of carbon-14 and tritium in toluene–PPO is the outer curve shown in Fig. 44. This curve shows carbon-14 being counted at about 60% efficiency with only 0.01% counting efficiency for tritium. These figures tend to vary among individual counters. The maximum tritium efficiency for an air-quenched sample is the tangent of the outer curve with a line perpendicular to the x axis, a value in excess of 50%. The best tritium–carbon-14 ratio for double-label

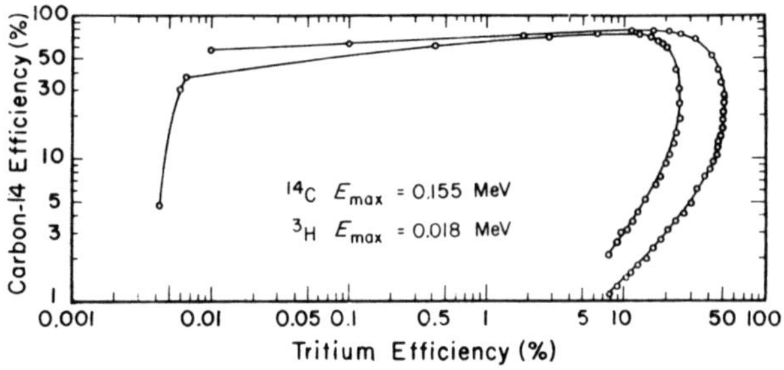

Fig. 44 Engberg plot for the isotope pair tritium and carbon-14. The inner curve was determined for a [^{14}C]toluene sample quenched with acetone from 72 to 32% counting efficiency as determined by the sample channels ratio method. The outer curve was determined using air-quenched samples of labeled toluene.

counting is the tangent to a 45° line drawn from the base of the curve. The 45° line represents a constant ratio between the carbon-14 and tritium counting efficiencies and, therefore, an equivalent rate of change of counting efficiencies. The upper portion of the curve deviating from the 45° line represents the best compromise between isotope ratios because this point indicates the highest tritium efficiency with the best tritium to carbon-14 ratio. Above this point, higher tritium efficiencies are obtained at the price of a less favorable isotope ratio.

The inner curve of Fig. 44 has been derived from a carbon-14 sample quenched from 72 to 32% counting efficiency by the addition of acetone. An equivalent amount of acetone has also been added to the tritium sample.

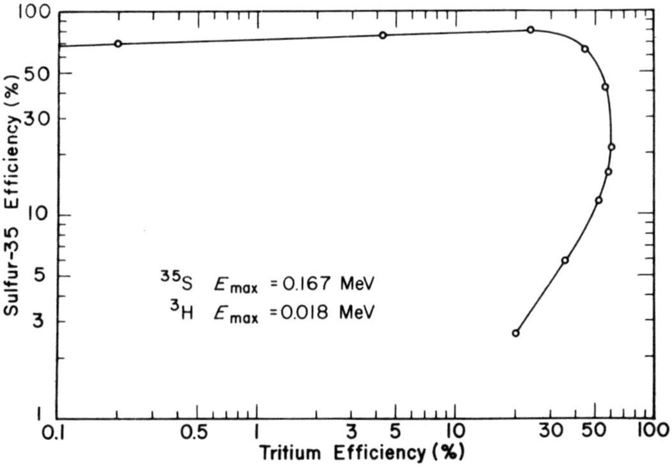

Fig. 45 Engberg plot for the isotope pair tritium and sulfur-35.

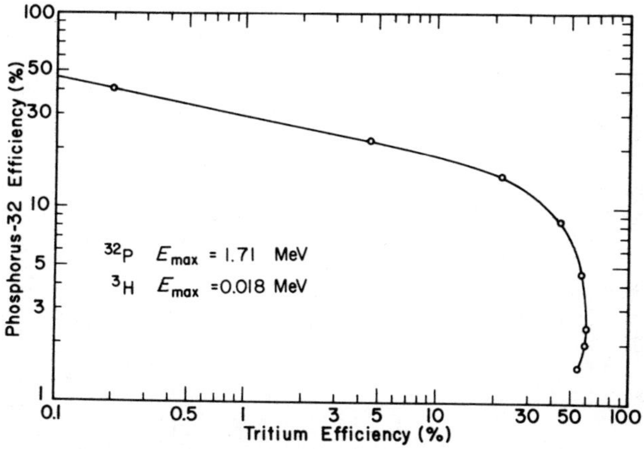

Fig. 46 Engberg plot for the isotope pair tritium and phosphorus-32.

The curve shows that the counting efficiency for carbon-14 can be restored almost completely by adjusting the amplifier gain. In contrast, the tritium efficiency can be restored to only half of its unquenched counting efficiency. The isotope efficiency ratio in the tritium channel has therefore deteriorated.

The utility of Engberg plots is that a single curve describes the counting characteristics for a given counter and any single type of counting solution. From this curve, one can determine at a glance the maximum counting efficiencies for both carbon-14 and tritium as well as the relationship between the counting efficiencies of the isotope pair for a given window width.

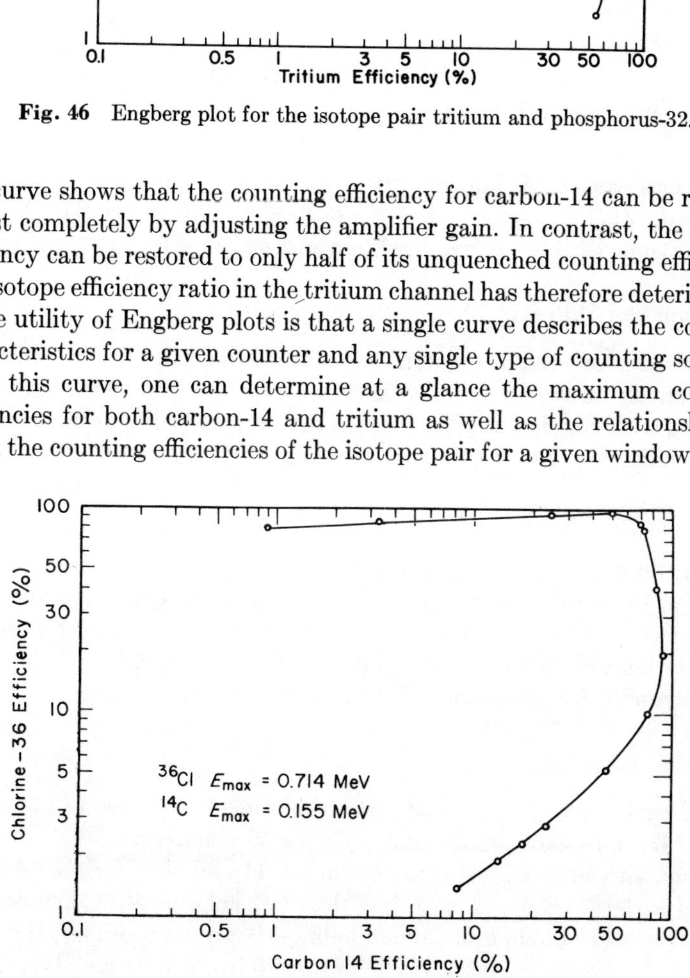

Fig. 47 Engberg plot for the isotope pair carbon-14 and chlorine-36.

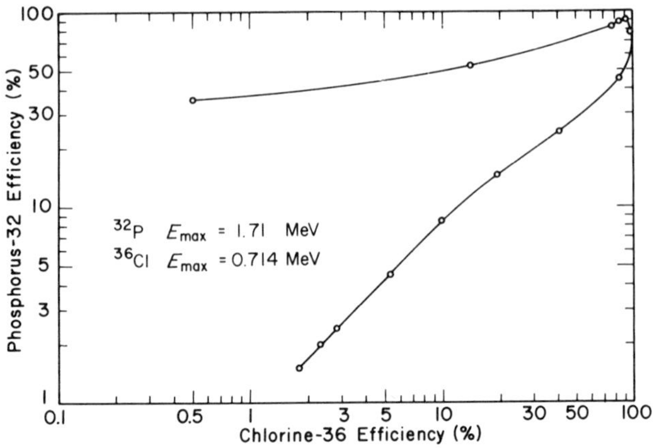

Fig. 48 Engberg plot for the isotope pair chlorine-36 and phosphorus-32.

Although carbon-14 and tritium is the most common isotope pair encountered, other pairs can and are also used to advantage. The Engberg plots for some of these pairs are shown in Figs. 45–48.

Tritium and Sulfur-35

Figure 45 shows the Engberg plot for the isotope pair tritium and sulfur-35. In an air-quenched sample, sulfur-35 can be counted at about 70% efficiency with less than 0.1% tritium contribution. Tritium can be counted at 50% efficiency with a sulfur-35 counting efficiency of 10%.

Tritium and Phosphorus-32

Figure 46 depicts the plot for the isotope pair tritium and phosphorus-32. In an air-quenched sample, phosphorus-32 can be counted at 90% efficiency with no tritium contribution. Tritium can be counted at 60% efficiency with a phosphorus-32 counting efficiency of only 2%. The 90% efficiency point for phosphorus-32 is not included in the figure.

Carbon-14 and Chlorine-36

The Engberg plot for the combination chlorine-36 and carbon-14 is shown in Fig. 47. This plot shows that chlorine-36 can be counted with high efficiency with little contribution of carbon-14 and that carbon-14 can be counted at 50% efficiency with less than 6% chlorine-36. Coincidentally, the beta energies of chlorine-36 and iodine-131 are very similar, the actual values of the two iodine betas being 0.61 and 0.81 MeV. It may be expected that the combination of carbon-14 and iodine-131 can be counted effectively in a liquid scintillation counter.

Chlorine-36 and Phosphorus-32

The last Engberg plot is for the pair chlorine-36 and phosphorus-32, shown in Fig. 48. A point to notice is that the ratio of the beta energies between the two isotopes is 2.4. About 10 years ago, isotope pairs with beta energy ratios of less than 4 or 5 could not be adequately analyzed with a liquid scintillation counter. However, the Engberg plot for this pair, chlorine-36 and phosphorus-32, shows that this is no longer true. Phosphorus-32 can be counted at 35% efficiency with only a 0.5% efficiency for chlorine-36, whereas chlorine-36 can be counted at 80% counting efficiency in the presence of 40% of the phosphorus-32. Although not an ideal counting situation, this is a usable system.

TESTING COUNTERS BASED ON LOGARITHMIC PULSE HEIGHT ANALYSIS

General Considerations

The basic difference between the counters with linear and logarithmic amplification is the lack of a gain or attenuator control in the logarithmic instruments. In these counters, the discriminator windows, which are adjustable from 0 to 100 divisions, cover the energy range of beta-emitting radioisotopes generally encountered in biological research, as illustrated in Fig. 49A. This illustrates that for tritium counting, the lower discriminator can be set at zero and the upper discriminator at 53 for maximum tritium counting efficiency. Similarly, the discriminators can be set between 0 and 68 for maximum carbon-14 counting efficiency, although very little of the spectrum lies below 30 divisions. By setting the lower discriminator at 30 divisions, therefore, very little counting efficiency for carbon-14 is lost. Moreover, most of the noise arising from photomultiplier tubes appears as background counts in the region of the tritium and carbon-14 spectra and so the elimination of the region between 0 and 30 divisions can result in a significant reduction in background. For counting low-activity samples this is often important. The value of raising the lower window for the purpose of reducing the background counts without significantly affecting the isotope counting efficiency is better illustrated by studying the chlorine-36 spectrum, Fig. 49A. Less than 5% of the chlorine-36 spectrum lies below

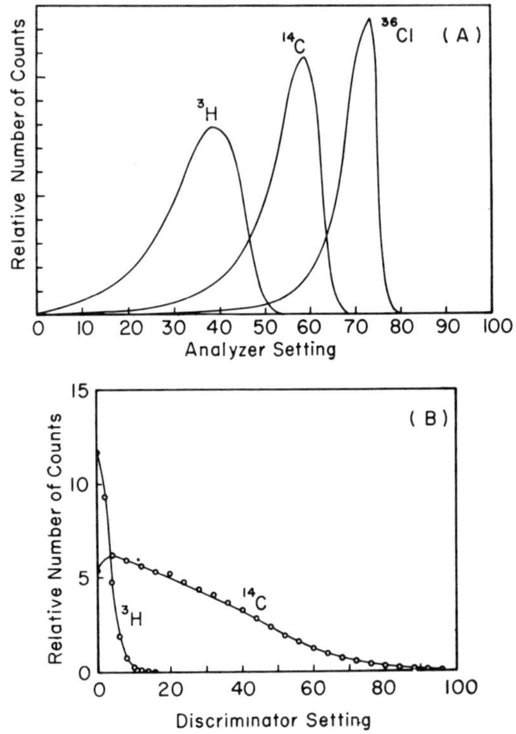

Fig. 49 (A) Spectra of tritium, carbon-14, and chlorine-36 determined in a liquid scintillation counter with logarithmic amplification. (B) Spectra of tritium and carbon-14 determined in a liquid scintillation counter with linear amplification.

50 divisions and the proper settings for efficiently counting this isotope are between 50 and 79 divisions. This includes 95% of the chlorine-36 spectrum.

In testing counters with logarithmic amplification, the major factors to consider are (a) the desired counting efficiency, (b) the background characteristics, and (c) the stability and reproducibility of the discriminators. In counting samples containing a single radioisotope, the reproducibility of the discriminators is not as critical as in the counting of doubly labeled samples. When the counter is set up for simultaneous counting of carbon-14 and tritium, the position of the upper discriminator for tritium and the lower discriminator for carbon-14 is critical. Study of the spectra of both radioisotopes shows that in the case of the upper discriminator for tritium, small changes in the region between 30 and 40 divisions result in a relatively large change in tritium counting efficiency, a feature that also applies to the lower discriminator setting for carbon-14. This is a consequence of the logarithmic system, because the spectra are compressed (on a log scale) when compared to those obtained when linear amplification is used. In a system employing linear amplification the spectrum can be spread over the entire width of the discriminators. This is illustrated in Fig. 49B. Assume that the spectrum is uniform over the 100 divisions of the discriminator. A change of one division is then equivalent to a 1% change in counting efficiency. Consider now the log spectrum of carbon-14. The entire spectrum lies between 18 and 68 divisions which is half the full width of the disscriminator. A change in one division is now equivalent to a 2% change in counting efficiency. In practice, the rate of change of counting efficiency per unit change of the discriminator setting of the log scale is larger for the upper discriminator of the tritium channel and the lower discriminator of the carbon-14 channel in the double-isotope counting mode because these positions correspond to the peak areas of their respective spectra.

A basic test for the stability of a log counter is to check it under modified conditions for double-isotope counting of tritium and carbon-14. The carbon-14 channel is set to accept approximately 1% of the tritium spectrum (about 1% counting efficiency for an unquenched tritium sample) and the tritium channel is set to accept approximately 6% of the carbon-14 spectrum. If the counter responds similarly to that used to plot Fig. 49A the carbon-14 discriminator settings are roughly 51–68 divisions and the tritium settings are about 0–36 divisions.

When testing the performance characteristics for counting a single isotope, the counter should be able to count an unquenched tritium sample at 60 ± 5% efficiency with an E^2/B figure of merit of better than 150. For an unquenched carbon-14 sample, the counter should be able to count in excess of 90% efficiency with a figure of merit greater than 300.

Discriminator Tracking Test for Logarithmic Units

The procedure outlined here is similar to that given for linear units but three different isotopes (tritium, carbon-14, and chlorine-36) are used to check the discriminators. For the sake of discussion, the discriminators are taken to have a range of 0–100 units. If the reference solutions contain approximately a 100,000 dpm or more, 1-minute counts are sufficient for this test. The discriminators are varied according to the schedule given in Table 10, and the reference isotope required is determined by the energy region being tested. The variable discriminators should track within 1% of each other. It may not be possible to check the region above 80 divisions in some counters with chlorine-36. However, if phosphorus-32 is available, more of the upper end of the discriminator can be checked for linearity because phosphorus-32 is more energetic than is chlorine-36.

TABLE 10

Discriminator Tracking Test Schedule for Log Units

Trial	Isotope	Lower discriminator	Upper discriminator
1	Tritium	0	5
2	Tritium	5	10
3	Tritium	10	15
4	Tritium	15	20
5	Tritium	25	30
6	Tritium	35	40
7	Carbon-14	40	45
8	Carbon-14	45	50
9	Carbon-14	55	60
10	Chlorine-36	65	70
11	Chlorine-36	70	75
12	Chlorine-36	75	80
13	Chlorine-36	80	90
14	Chlorine-36	90	100

PLUG-IN DISCRIMINATORS

Many new counters now have preset plug-in or pushbutton discriminators that are not adjustable by the operator. The main requirement is to check each plug-in unit for equal counting characteristics in all counting stations. Each unit should count an appropriate reference radioisotope within 1% of each other at all counting stations. Electrically speaking, discriminators are precision resistors and when they are plugged in and out of counting channels, it is essential that the contacts be kept absolutely clean. Fingerprints, for example, can change the resistance of the contact. Similarly, the plug-in discriminators not in use should be kept in a noncorrosive atmosphere. Excessive changing of these plug-in units should be avoided because the electrical contacts become worn, resulting in a change in the resistance of the contacts and a consequent shift of the discriminator window.

Pushbutton discriminators are similar to plug-in discriminators except that they are permanently wired into each counting channel.

THE CHI-SQUARE TEST AND DETERMINATION OF COUNTER STABILITY

When a counter is being examined for statistical counting behavior, a Poisson distribution is assumed to be normal and Pearson's chi-square test is applied. The purpose of this test is to establish the probability, P, that repeated counts on a given sample fall within established limits of variability. The test is independent of the actual count rate.

Chi-square is defined as follows.

$$\chi^2 = \sum_n \frac{[(\text{observed value})n - (\text{average value})]^2}{(\text{average value})}$$

where n = numbers of observations and the average value is the mean value of all the observations. (The average value is assumed to be the expected value.)

In order to determine any statistics, the number of degrees of freedom, F must be known. For a Poisson distribution

$$F = n - 1$$

The probability P can now be determined by consulting any standard statistical handbook. The most desirable value of P is 0.5. In counting statistics, the acceptable limits have been arbitrarily set by Jarrett (1) to lie between P values of 0.1 and 0.9. Low values of P indicate high variability and may indicate instrument instability. High values of P suggest the lack of variability or a nonrandom behavior of the counter. High P values

TABLE 11

Statistical Analysis of Counting Data for Instrument Evaluation[a]

Trial	X	$X - \bar{X}$	$(X - \bar{X})^2$
1	10,800	−54	2,916
2	10,900	46	2,116
3	10,750	−104	10,816
4	11,000	146	21,316
5	10,820	−34	1,156
Totals, Σ	54,270	0	38,320

[a] Number of observations = N = 5.
Degree of freedom = $N - 1 = (5 - 1) = 4$.
Average counts = $\bar{X} = \Sigma X/N$ = 54,270/5 = 10,854.
Calculated standard deviation = $\sigma = \sqrt{\bar{X}} = \sqrt{10,854}$ = ±104.18.
Experimental standard deviation = $\sigma = \sqrt{\Sigma(X - X)^2/N - 1} = \sqrt{38,320/5 - 1}$ = ±97.87.
Standard error of the Mean = Experimental standard deviation/$\sqrt{N}$ = 97.87/$\sqrt{5}$ = ±43.88.
Chi-square = $\chi^2 = \Sigma(X - \bar{X})^2/\bar{X}$ = 38,320/10,854 = 3.53.
Probability of χ^2 value for five observations = P = 0.47.

can be caused by the generation of a constant and high level of spurious "noise" that is appearing as counts.

An example of a chi-square test is shown in Table 11. In the interest of clarity, only five counts are analyzed. Normally, a minimum of 20 determinations are made. The acceptable ranges of chi-square values for 5–40 determinations are given in Table 12. Note that the counts analyzed represent the total number of counts accumulated during the counting period. For example, in trial 1 of Table 11, the first count of 10,800 may have been accumulated in 1, 5, 10, or 30 minutes. If the counts have been accumulated in 5 minutes, it is statistically incorrect to calculate the counts per minute and then to analyze the computed data for χ^2 distribution. In computing the counts per minute, the data become biased in favor of high P values because some of the original variability is averaged out. If these counts have been obtained in 5 minutes, then the results are $10{,}584/5 \pm 97.87/5$ counts or 2117 ± 19.57 cpm (mean ± standard deviation).

The first test for counter stability is made at the counting conditions determined earlier for counting tritium and carbon-14 singly at high efficiency (at the balance point in linear counters, and at the highest efficiency in log counters). Sealed reference sources of unquenched carbon-14, tritium, and a background sample are counted consecutively for 10-minute periods and repeated a minimum of 20 times. Loading and unloading the sample between counts equalizes the variations in geometry between the sample vial and the photomultiplier tubes. The counter should easily meet the chi-square criterion during single-isotope counting of any radioisotope.

A more stringent test for stability is to evaluate the counter under double-isotope counting conditions, in which the counter is operated off balance point. In this test, the lower window of the carbon-14 channel is adjusted

TABLE 12

Chi-Square Limits (P = 0.9–0.1)

Number of determinations	Chi-square limits
5	1.064–7.779
10	4.168–14.684
15	7.790–21.064
20	11.651–27.204
25	15.659–33.196
30	19.768–39.087
35	23.952–44.903
40	28.196–50.660

to include about 1% of the unquenched tritium standard. The same three samples used to test the single-isotope settings are again employed. Twenty 10-minute counts are accumulated for each of the three samples and the χ^2 analysis is done for (a) carbon-14 counts in the tritium channel and (b) the tritium counts in the carbon-14 channel. A stable counter should pass the χ^2 test. If the tests are satisfactory for these two determinations, they are usually satisfactory for tritium in the tritium channel and carbon-14 in the carbon-14 channel. Any tendency of the instrument to "drift" is indicated by either a gradual decline or an increase in counts in one channel with the opposite change occurring in the other channel. The chi-square analysis of these data shows the statistical behavior of the counter. Unfortunately, a counter can meet the statistical requirements of stability and reproducibility and still be unacceptable. This situation arises when the counter performance slowly changes so that a 24-hour period is not sufficient to monitor this change. These changes become apparent over a period of time and can cause a serious error when a number of low-level activity samples are counted over several days. An example of such a case is shown in Table 13. On a daily basis the chi-square test is satisfactory, but the average value of the sealed standard changes significantly from Monday to Friday.

In practice the counting characteristics of a counter may not become constant for about 6 weeks after installation but the changes occurring after the first week are usually minor and should not unduly influence counting statistics. A simple method of determining whether or not two counts on the same sample are statistically different (because of instrument variation) has been suggested by Jarrett (1). This method consists of comparing the observed difference between the two determinations, N_1 and N_2, and the standard error of the difference between the two determinations.

TABLE 13

Chi-Square Test on a Sealed Carbon-14 Reference Source during One Week

	Monday	Tuesday	Friday
N (observations)	20	20	20
Time per observation (minutes)	10	10	10
Average total counts	742,645	741,675	739,305
Calculated standard deviation	861.7	861.2	859.8
$\Sigma(X - \bar{X})^2$	12,100,000	19,800,000	9,800,000
Experimental standard deviation	798.0	1,020.8	718.2
Chi-square	16.29	26.69	13.25
P	0.6–0.7	0.1–0.2	0.8–0.9

The standard error of the difference is given by the formula

$$Y = \sqrt{\frac{N_1}{t_1} + \frac{N_2}{t_2}}$$

where N_1 = net counts per minute of first count
t_1 = total counting time of first count in minutes
N_2 = net counts per minute of second count
t_2 = total counting time of second count in minutes

The probability P that N_1 and N_2 are different is determined by calculating the ratio R of the observed difference to the standard error of the difference $(N_1 - N_2)/Y$ and finding the probability in Table 14.

In counting, significance is arbitrarily ascribed to P values of 0.01 or less. This means that the probability must be less than one chance in 100 for the observed difference to be considered caused by something other than the statistics of isotope decay.

When this determination is applied to the average counts observed on Monday and Friday of Table 13, the following results.

$$Y = \sqrt{\frac{N_1}{t_1} + \frac{N_2}{t_2}} = \sqrt{\frac{74{,}264}{10} + \frac{73{,}931}{10}}$$

$$Y = \sqrt{14{,}819.5} = 121.73 \text{ cpm}$$
$$R = (N_1 - N_2)/Y = (74{,}264 - 73{,}931)/121.73$$
$$R = (334/121.73) = 2.73$$

A value of 2.7 for R in Table 14 gives a probability of 0.004. This means there are four chances in 1000 that this difference has occurred by chance. As the arbitrary standard of significance is 0.01, it can be surmised that the instrument is not stable.

TABLE 14

Probability P of Observing $(N_1 - N_2)/Y$ Equal to or Greater than R[a]

R	P	R	P	R	P
1.0	0.159	2.0	0.023	3.0	0.0013
1.2	0.115	2.2	0.014	3.2	0.00068
1.4	0.081	2.4	0.008	3.4	0.00034
1.6	0.055	2.6	0.005	3.6	0.00016
1.8	0.036	2.8	0.003	3.8	0.00007

[a] R less than 1.0 has no significance.

Long-term drift is a common problem in new liquid scintillation counters. A daily log of their counting characteristics on sealed reference standards is the best way to check on their stability. Any undue amount of drift can be easily spotted by an inspection of such a counter log.

DETERMINATION OF COUNTING EFFICIENCY

The main objective of any counting procedure, be it gamma counting for an immunoassay procedure, soft beta counting of a solid sample using a Geiger counter, or soft beta counting in a liquid scintillation counter, is the accurate determination of the radioactive content of the sample. In liquid scintillation counting, this cannot be accomplished without first determining the efficiency with which the radioactive content of the sample is assessed. Counting efficiency is commonly determined by one of three methods: (a) by internal standardization, (b) by the channels ratio method or (c) by the external standard channels ratio method. The relative merits of these methods of efficiency determination have already been discussed.

Internal Standard Method

A sample containing either tritium or carbon-14 is first counted. After the count, a small amount of standardized radioactive toluene containing about 50,000 dpm of carbon-14 or 150,000 dpm of tritium in a volume of 0.1 ml or less is added to the sample and the sample is recounted for 1 minute. The internal standard activity must always be greater than the sample activity for statistical reasons. The counting efficiency is equal to the additional counts caused by the standard divided by the value of the standard in disintegrations per minute multiplied by 100.

If the sample is removed from the counter, the sample should be allowed to thermally equilibrate before it is opened to prevent moisture from condensing in the cold sample vial. Great care must be exercised in adding the internal standard if the standardization is to be accurate.

The internal standard method is the method of choice in many double-isotope methods, especially where low sample activity is commonplace. All samples are similarly prepared and assumed to count at the same efficiency. The count rates of carbon-14 alone in the carbon-14 and tritium channels are determined experimentally using a standardized carbon-14 sample containing about 50,000 dpm. The count rates for tritium alone are also determined in the two counting channels. The count rates found for these two standards are then used to determine the constants for the double isotope equation derived in Chapter 2, page 23. An example is given below using both the regular double isotope equation and the simplified form of the equation (isotope exclusion method) for the same data. It is apparent that the small error introduced by the simplified equation is negligible when very small amounts of tritium are accepted into the carbon-14 channel.

Example

Amount of standard carbon-14 used	50,000 dpm
Amount of standard tritium used	200,000 dpm
Background in tritium channel	21 cpm
Background in carbon-14 channel	16 cpm

	Gross counts (cpm)	Efficiency (100% = 1)	
Gross carbon-14 counts in carbon-14 channel	30,016	0.60	(c_1)
Gross carbon-14 counts in tritium channel	3,521	0.07	(c_2)
Gross tritium counts in carbon-14 channel	66	0.00025	(h_1)
Gross tritium counts in tritium channel	110,021	0.55	(h_2)
Net cpm in channel A (carbon-14 channel)			(N_1)
Net cpm in channel B (tritium channel)			(N_2)

By simultaneous equation

$$C = \frac{N_1 - N_2(h_1/h_2)}{c_1 - c_2(h_1/h_2)} = \frac{N_1 - N_2(0.00025/0.55)}{0.6 - 0.07(0.00025/0.55)} = \frac{N_1 - 0.000454N_2}{0.599968}$$

$$H = \frac{N_2 - N_1(c_2/c_1)}{h_2 - h_1(c_2/c_1)} = \frac{N_2 - N_1(0.07/0.6)}{0.55 - 0.00025(0.07/0.6)} = \frac{N_2 - 0.116666N_1}{0.549947}$$

Data for unknown sample

Gross counts in carbon-14 channel = 1016 cpm
Gross counts in tritium channel = 3021 cpm

Solution for unknown by simultaneous equation

$$C = \frac{(1016 - 16) - 0.000454(3021 - 21)}{0.599968}$$

$$= \frac{1000 - 1.36}{0.599968} = \frac{998.64}{0.599968} = 1664.48 \text{ dpm}$$

$$H = \frac{(3021 - 21) - (1016 - 16)(0.11666)}{0.549947}$$

$$= \frac{2883.33}{0.549947} = 5242.92 \text{ dpm}$$

Solution for unknown by simplified equation

$$C = N_1/c_1$$

$$H = (N_2 - Cc_2)/h_2$$

$$C = (1016 - 16)/0.60 = 1666.66 \text{ dpm}$$

$$H = \frac{(3021 - 21) - (1666.66)(0.07)}{0.55} = \frac{(3000) - (116.66)}{0.55}$$

$$H = 5242.42 \text{ dpm}$$

Sample Channels Ratio Method

The sample channels ratio method is based on a mathematical relationship derived by plotting the change in the ratio of the sample count rate as it is altered by quenching in two counting channels the discriminators of which are set to see different portions of the same beta spectrum. The accuracy of the channels ratio plot depends on the accuracy of the series of quenched standards used to plot the curve. It is also convenient to establish a curve that is linear in the region where the experimental samples usually fall. In most experiments, the samples are prepared in a manner that ensures a narrow range of quenching; that is, the samples have a uniform counting efficiency (within 0.5% of each other). For this reason, preparing a quench curve covering a wide range of quench, for example, from 5 to 60% counting efficiency for tritium, is seldom useful or necessary. With good sample preparation, a range between 20 and 60% for tritium should be more than adequate and a range of 40–80% adequate for carbon-14. When samples are severely quenched, it is better to change the counting conditions, either by changing the method of sample preparation to reduce quenching or by changing the counting conditions of the counter by raising the gain (or reducing attenuation) to restore the quenched spectrum back into the discriminator window. In logarithmic units, the latter can be accomplished by moving the window toward the lower pulse height region.

A satisfactory quench curve for tritium can be constructed by using a wide window, such as 2–100 divisions, for channel A at the tritium balance point and a narrow window (2–30 divisions) for channel B. Both channels A and B are maintained at the same gain (or attenuation) setting. Channel B should be set to accept approximately 30–35% of the channel A count rate. In counters with logarithmic amplification, channel A should take in the entire tritium spectrum and channel B about 30–35% of the lower portion of the tritium spectrum. By plotting the ratio channel B/channel A against the counting efficiency found for tritium in channel A, a quench curve that is almost linear should be obtained. A typical quench curve for tritium is shown in Fig. 50. Quench curves derived from commercially prepared quenched standards are not applicable to determining the counting efficiency of the experimental samples unless the composition of the counting solution is identical to that of the samples. This is because the counting efficiency is a function of the scintillator, its concentration, and the solvent used.

A convenient way to construct a quench curve for experimental samples is illustrated by the following example. Dispense 10 ml of a toluene–PPO counting solution into a counting vial and add approximately 200,000 dpm of tritiated toluene (in about 0.1 ml). Count the sample and record the counts in channels A and B, which are set up as described above (channel A = maximum tritium efficiency, channel B = approximately one-third the count rate of channel A). Then add 0.05 ml of acetone and recount. Repeat this procedure until 0.5 ml of acetone has been added. This procedure provides data for a ten-point quench curve covering a range of quench adequate for most purposes. However, if a wider range curve is

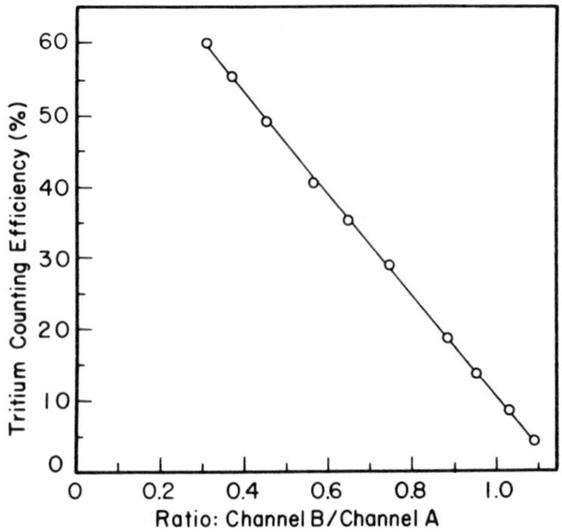

Fig. 50 Sample channels ratio versus tritium counting efficiency plot. Channel A is used for sample analysis and channel B for the ratio channel B/channel A.

desired, nitromethane can be substituted for acetone as the quencher. The accuracy of the addition of the quencher (acetone or nitromethane) is not critical. The volume of the quenching agent added to the counting solution should be less than 0.5 ml to avoid a significant dilution of the scintillator concentration.

Although the samples channels ratio method of efficiency determination has been suggested for double-labeled samples, the method is only useful when the sample contains a significant amount of the higher energy isotope. This permits an accurate channels ratio analysis using only the spectrum of the higher energy isotope above that of the lower energy isotope. This single ratio is then related empirically to (a) the efficiency of the high-energy isotope in the high-energy isotope counting channel, (b) the counting efficiency of the lower energy isotope in the lower energy isotope counting channel, and (c) the counting efficiency of the high-energy isotope in the lower energy isotope counting channel. A means of choosing the best conditions for counting for double-labeled samples has been mathematically determined by Bush (2, 3). If the samples channels ratio is to be used effectively for double-labeled samples, all samples must contain a sufficient amount of the higher energy isotope to allow an accurate determination of the sample channels ratio within a reasonable counting time and, second, the sample must not be so severely quenched that the higher energy spectrum is shifted out of the discriminator window.

External Standard Method

The use of the external standard channels ratio for the determination of counting efficiency in singly labeled samples is not recommended unless the samples are of relatively low activity. The samples channels ratio is a more direct and accurate means of efficiency determination. However, the external standard method is useful for double-labeled samples because it is independent of the type and the amount of isotope contained in the sample.

To determine the amount of each radioisotope contained in a double-labeled sample, three curves are needed. If the two isotopes being studied are carbon-14 and tritium, the curves required are

1. The external standard ratio versus carbon-14 efficiency in the carbon-14 channel
2. The external standard ratio versus tritium efficiency in the tritium channel
3. The external standard ratio versus carbon-14 efficiency in the tritium channel

It is assumed that the lower energy isotope, tritium, is excluded from the carbon-14 channel and that the tritium channel has been optimized for best double-isotope counting statistics. A series of variably quenched carbon-14 samples are now counted with and without the external standard. The external standard count ratio is then plotted against the carbon-14 efficiencies in the two sample counting channels. A series of variably quenched tritium samples is now counted with and without the external standard. If the gamma source used is barium-133, the beta spectrum of the Compton electrons generated by irradiation of the sample is similar to that of carbon-14. The counts contained in the sample itself therefore contribute to the total counts collected in the two external standard counting channels during irradiation of the sample by barium-133. With barium-133 the sample is first counted in the usual counting channels. The sample is then counted for a short period in the two external standard counting channels in the absence of barium-133 and, finally, in the external standard counting channels in the presence of barium-133. The difference between the last two counts is caused exclusively by the external standard and these results are used to compute the external standard channels ratio. A series of experimental correction curves obtained using the external standard channels ratio is shown in Fig. 51.

In some older counters, the counts caused by the external standard are monitored in a single counting channel set to count only very energetic Compton electrons generated by a relatively high-energy gamma source. These counts are then related directly to counting efficiency. However, this

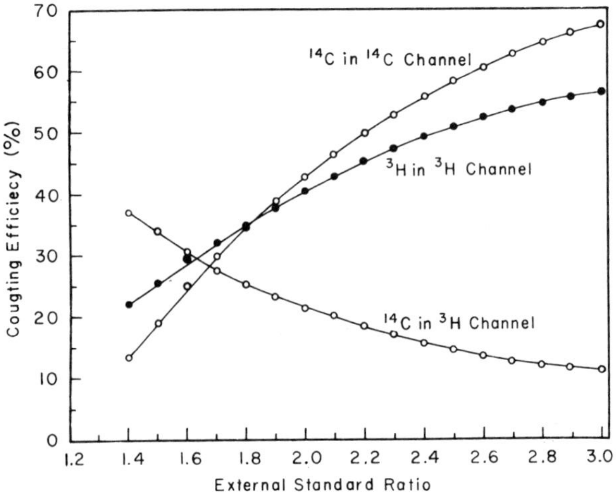

Fig. 51 External standard channels ratio plots for counting efficiency determination of doubly labeled samples (carbon-14 and tritium).

first approach to the use of an external standard has been found to be unsuitable, because of its volume dependency and its insensitivity to low degrees of quench, and is therefore not recommended.

EFFICIENCY DETERMINATION IN HETEROGENEOUS COUNTING SYSTEMS

A problem that is often ignored is the determination of counting efficiency in heterogeneous systems. The basic problem is that with experimental samples, there is no way the investigator can determine with certainty the partition coefficient of the radioactive material between the various phases of a heterogeneous counting mixture. For example, when radioactivity is counted on a piece of filter paper, partial elution of the radioactivity results in different counting efficiencies for the soluble and insoluble portions, the soluble fraction counting with 4π geometry with no self-absorption losses and the insoluble fraction counting at 2π geometry together with some possible self-absorption losses. A similar situation arises when adsorption of the radioactive material occurs on the vial wall.

In an emulsion system, numerous examples of the inability to determine counting efficiency by either the sample or external standard channels ratio have appeared in the literature. This is because in emulsions and gels the partitioning of the radioactivity between the aqueous micelles and the organic solvent is variable. It has been observed that efficiency determination using two different internal standards, [^{14}C]toluene or a [^{14}C]amino acid, can give two different values in an emulsion system (4, 5). This phenomenon is brought about by the greater solubility of toluene in the organic phase; because this contains the most of the scintillator the toluene is counted at higher efficiency compared to the amino acid, which being less soluble is mainly in the aqueous phase where the emitted beta particles must penetrate the aqueous micelle to interact with the scintillator in the organic phase. The latter process also results in some quenching. This problem is recognized by Bush (6), who has studied the potential of using either the sample or external standard channels ratio technique to determine counting efficiencies in heterogeneous counting systems. She has found that no procedure is reliable, including the use of an internal standard. When phasing occurs, although convincing evidence may be presented that no radioactivity is present in one phase, the possibility always remains that spurious optical effects introduced by the second phase contribute to an error in efficiency determination. The effects of spectral distortions produced by haziness, opalescence, and light scattering by fine suspended particles are also erratic and unpredictable, leading to errors in sample counting efficiency determination. The latter effects are diminished with increasing beta energy. Bush has found that when both the samples channels ratio and the external standard channels ratio are used to determine the sample counting efficiency for a homogeneous solution, a curve-linear relationship is found (Fig. 52) because they both measure solution counting efficiency. A disparity between the two channel ratio values gives points that are not on the curve and is indicative of samples that are not true solutions.

The external standard channels ratio measures only the solution counting efficiency, whereas the samples channels ratio measures both the solution counting efficiency and the extent to which the radioactive sample is in solution. When the radioactive sample is only partially dissolved, the samples channel ratio appears more quenched, whereas the external standard ratio does not alter because the solution counting efficiency remains unchanged. The upper curve of Fig. 52 shows the plot of the samples channel ratio versus counting efficiency. The point at the ratio of 0.398 represents the only case found by Bush (6), among 150 samples tested, in which the

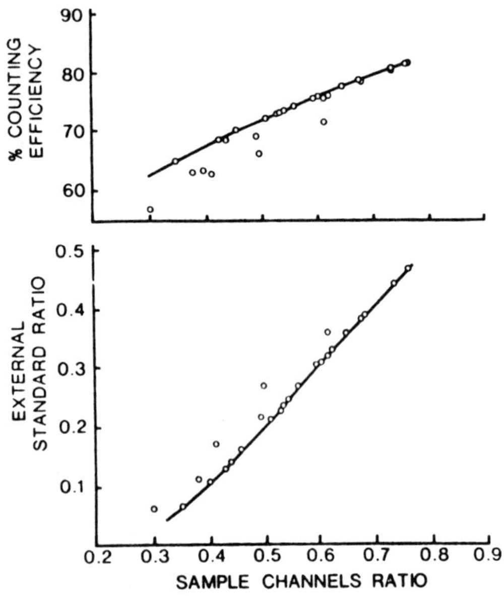

Fig. 52 A double-ratio plot to determine partial solubility of radioactive material (^{35}S) in a counting solution. The upper curve is a plot of the counting efficiency versus the sample channels ratio. The lower curve is the plot of the external standard ratio versus the samples channels ratio. Homogeneous samples are those that fall on the curve and partially soluble samples are those that are off the curve (6).

double-ratio plot has failed to identify an errant sample. Confirmation of the double-ratio technique has been provided by Houx (7), who has found it successful in detecting scrapings from thin-layer plates that have only partially been eluted by the counting solution. He has also reported that the external standard ratio cannot be used to determine counting efficiency of silica gel-containing samples in which complete elution has occurred, whereas the samples channels ratio is satisfactory.

The double-ratio plot is a useful qualitative tool to determine whether a sample is completely dissolved in a counting solution. Bush has found no useful, predictable relationship between the position of an errant value on a double-ratio plot and the various parameters, such as adsorption, precipitation, self-adsorption, or back scattering. This study points out the difficulty of accurately determining the counting efficiency of some heterogeneous counting systems. If heterogeneous counting systems are employed, great care must be exercised to ensure that the counting efficiencies can be assessed with confidence.

CURVE FITTING

The most commonly used method of curve fitting for quench correction purposes is the least-squares fit to a polynomial of the form

$$Y = a + bx + cx^2 + dx^3 + \cdots + nx^z$$

where Y = the counting efficiency
X = the channels ratio

and the constants $a, b, c, \ldots, n$ are generated by a mathematical procedure.

The principal advantages of the least-squares method are that (a) it provides the best statistical fit of the data which describes an easily interpreted smooth curve and (b) the method is completely independent of any subjective considerations and the results can be applied to any channels ratio between the highest and lowest values used to generate the equation for determining counting efficiency.

The major weakness of the least-squares method is that there is no allowance for judgment as to the suitability of all the data points used in the analysis. The method automatically gives equal weight to all data points with the consequence that "bad" points can unduly influence the final result. It is usually advisable to roughly plot the data by hand and eliminate any obviously poor points before computation.

When the least-squares program is executed, the procedure is to start from the lowest degree polynomial that may fit the data and to keep increasing the degree of the polynomial until an acceptable degree of fit is obtained. For example, an index of determination is often given for each analysis. The equivalent statistical term for the index of determination is the multiple correlation coefficient squared. An index of 1.0 indicates a perfect fit, but an index of 0.998 or better is acceptable. For most quench curves, this index is obtained by a third- or fourth-order polynomial. The lowest degree polynomial that achieves an index of determination of 0.998 is used for the quench curve.

A cautionary note is offered in the application of the least-squares fit to a polynomial. A basic rule is that one should always have more data points than the degree of the polynomial, i.e., more than four points for a fourth-degree fit. Mathematically, one always arrives at a perfect fit when the degree of the polynomial is equal to the number of points. The perfect fit merely says that the curve described by the polynomial passes through every point. This is illustrated in the following example.

Table 15 gives a series of hypothetical data points for a quench curve known to be linear. The linear regression line, which is the first-order fit, and the eighth-degree polynomial, which gives a perfect fit, are shown in Fig. 53. It can be seen that although the eighth-degree polynomial is a better mathematical fit than the simple regression line, the regression line is a more accurate description of the quench curve. In these days of instant computations, it is easy to be lulled into a false sense of security by having "infallible" computerized data.

TABLE 15

Channels Ratio versus Efficiency Data for Quench Curve

Channels ratio	Efficiency
0	0
0.1	9.5
0.2	21
0.3	29
0.4	41
0.5	48
0.6	62
0.7	68
0.8	82

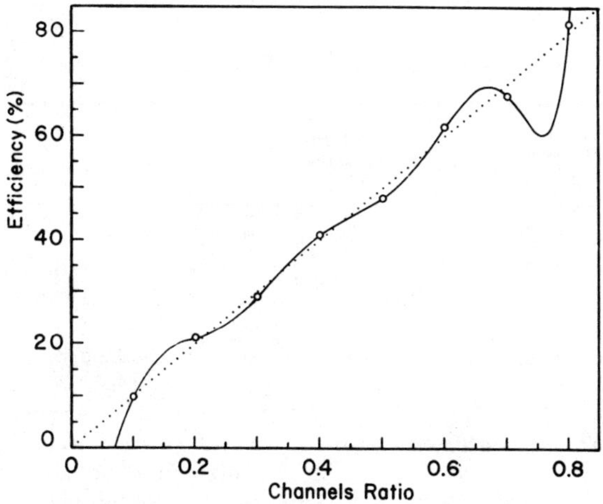

Fig. 53 Least-square fit to a polynomial for data given in Table 15. The dotted line is the regression line, efficiency = $-0.2444 + 100.75X$, with an index of determination of 0.9969. The solid line is the eighth-order fit, efficiency = $0 - 784X + 21{,}236X^2 - 194{,}886X^3 + 909{,}008X^4 - 2{,}365{,}906X^5 + 3{,}475{,}700X^6 - 2{,}691{,}472X^7 + 853{,}176X^8$, with an index of determation of 1.0.

COUNTING STATISTICS

The random nature of radioactive decay makes the determination of the "true" radioactive content impossible, although an estimate with an acceptable degree of error can always be made. In general, the larger the number of events observed, the greater the precision. In radioactive work, the statistical considerations are based on the assumption that the distribution of data follows the so-called "Poisson statistics" and the error e of any averaged value is related mathematically to a confidence factor k, multiplied by the square root of the total number of counts N collected.

$$e = k\sqrt{N}$$

When the confidence factor k is one, the error e is called the "standard deviation" (σ). In the biological sciences, confidence factors are considered as multiples of the standard deviation. An averaged value $\pm$ one standard deviation (1σ) includes 68.26% of all values drawn from the same population. Similarly, $\pm 2\sigma$ includes 95.46% of all values and $\pm 3\sigma$ includes 99.73% of all values. A listing of the commonly used error terms in statistics is contained in Table 16.

TABLE 16

Summary of Errors Commonly Used in Statistical Analysis

Name	Formula	Probability[a] (%)	Comment
Probable error	$\pm 0.65\sqrt{N}$	50	Although this error is commonly used, it is archaic from a statistical point of view; only 50% of the values are within the mean plus or minus the probable error
Standard deviation	$\pm\sqrt{N}$	31.74	The most commonly used error because many statistical tables are computed as functions of the standard deviation
Nine-tenth error	$\pm 1.645\sqrt{N}$	10	Popular in radio counting; called nine-tenth error because there are nine chances out of ten that the error is smaller; statistically referred to as the 0.1 level of significance
Ninety-five hundredth error	$\pm 1.96\sqrt{N}$	5	Most commonly used error by statisticians as the level at which deviations exceed chance variation; all values beyond the range of the mean $\pm$ error are considered significant (from a different population); called the 0.05 level of significance
Two times standard deviation	$\pm 2\sqrt{N}$	4.54	All errors listed below a probability of 5% are considered levels at which deviations exceed chance variation
Ninety-nine hundredth error	$\pm 2.576\sqrt{N}$	1	Values greater than two standard deviations from the mean are considered significant (different), whereas values greater than three standard deviations are highly significant
Three times standard deviation	$\pm 3\sqrt{N}$	0.27	

[a] Of observing an error equal to or greater than that error named in column one.

In all assays done by liquid scintillation counting, corrections must always be made for the background or the experimental blank value. This determination, as with any measurement, has an associated error that must be considered. In radioactive counting, the errors are summed according to the standard statistical procedure for the addition of independent errors, which is defined mathematically as

$$S_T = \sqrt{S_1^2 + S_2^2}$$

where S_T = total error

S_1 = error of background measurement

S_2 = error of sample count measurement

The S_1 and S_2 can be any error (probable, standard deviation, etc.) as long as they are the same class of error.

Example

A sample is counted for 10 minutes and 3600 counts are collected. A proper background sample counted 30 minutes gives 576 counts. The net sample count ± standard deviation is calculated as follows.

$$\text{Sample count rate} = \left(\frac{3600 \pm \sqrt{3600}}{10 \text{ minutes}}\right) = 360 \pm 6 \text{ cpm}$$

$$\text{Background} = \frac{576 \pm \sqrt{576}}{30 \text{ minutes}} = 19.2 \pm 0.8 \text{ cpm}$$

$$\text{Net sample count rate} = 360 - 19.2 \pm \sqrt{6^2 + 0.8^2} = 340.8 \pm \sqrt{36.64}$$

$$= 340.8 \pm 6.05 \text{ cpm}$$

Rejection of Counting Data

When a series of similar samples is counted, occasionally an unusual count is observed that is obviously different from the rest of the group. Experimentally and instrumentally, there is no obvious basis for this discrepant count and, therefore, no reason can be offered to disregard this one value. However, to include it can unduly influence the averaged value for the group. The canons of science dictate the inclusion of all data but to include a chance odd reading may significantly alter the conclusions drawn from these data. In order to deal with this dilemma, a reasoned basis for data rejection has been suggested by Chauvenet, and his procedure, known as "Chauvenet's criterion," has been the accepted means of rejecting unusual data.

Chauvenet's criterion states that any reading in a series may be rejected if the magnitude of its deviation from the mean of n values (including the suspect value) is such that the probability of occurrence of all deviations of this value or larger is less than $1/2n$. For example, if there are ten similar samples counted and one extreme value is suspect, the criterion for rejection is whether this value exceeds the mean ± the deviation equivalent to

$$p \leqslant \frac{1}{2n} = \frac{1}{2(10)} = \frac{1}{20} = 0.05$$

A p value of 0.05 of significance is reached by any value outside the range of the mean $\pm\ 1.96\sqrt{\text{mean}}$ or $\pm 1.96\sigma$. Therefore, according to Chauvenet's criterion, any of the ten values that is outside this range may be rejected. If a value is rejected, the mean and standard deviations are recomputed without the errant value. Table 17 gives Chauvenet's criteria for rejecting values for various number of observations.

Although Chauvenet's criterion has long been used as a basis of rejecting observations, especially in counting statistics, there are some problems associated with it. The main objection raised is that for a small number of observations, values are rejected too easily. However, a universal procedure for rejection of a suspect value from a given series remains unresolved. Some of the more recent attempts for the rejection of extreme values are given in the United States National Bureau of Standards Handbook 91, "Experimental Statistics" (8).

TABLE 17

Chauvenet's Criterion for Rejecting a Value[a]

n	k	n	k	n	k
2	1.15	7	1.80	20	2.24
3	1.38	8	1.86	25	2.33
4	1.54	9	1.91	30	2.40
5	1.65	10	1.96	35	2.45
6	1.73	15	2.13	40	2.50

[a] n is the number of observations and k is the factor that, when multiplied by the standard deviation, defines the range from the mean beyond which values may be rejected for that n.

AUTOMATED TECHNIQUES FOR SAMPLING COUNTING EFFICIENCY DETERMINATION

The rapid development of integrated circuits and compact memory cores has allowed the instrument manufacturers to incorporate data processing into the design of a liquid scintillation counter without increasing the physical dimensions of existing counters. Most commercial firms offer units with this capability but, for competitive reasons, use different approaches to reduce counting data to disintegrations per minute. The only common denominator among the various techniques is that all of the "instant" data options reduce the counting data of individual samples and cannot do any statistical evaluations of either repeated counts or groups of samples. Statistical analysis must be done separately or by a dedicated online computer. The use of a dedicated computer with the capacity to do all the necessary computations on-line for a single counter makes for inefficient utilization of computer time and is difficult to justify on economic grounds. To cope with this situation, most manufacturers are offering some system of multiplexing several counters to a single computer.

Packard Absolute Activity Analyzer

The Packard absolute activity analyzer, current model 544, is compatible with the Tri-carb liquid scintillation counter, model 3390, or its equivalent manufactured by the Packard Instrument Company.

The basic concept underlying the development of the absolute activity analyzer is the observation that the external standard ratio curve covering a wide range of quench prepared with many data points does not produce a smooth curve that can be described either by a simple regression line or by a second-degree polynomial of the form, $a + bx + cx^2$. Up to a fifth- or sixth-order polynomial is required to mathematically define such a curve with a small degree of error. To eliminate the error of interpolating between the various points of a quench curve, an instrument has been designed that can electronically alter the quench of a sample by manipulating the magnetic field between the photocathode and the first dynode of the photomultiplier tube precisely to one of ten external standard ratios varying from 0.9000 to 0 in steps of 0.1000, where the counting efficiencies have been previously determined with a set of ten calibrated, variably quenched standards. For a double-labeled sample, there are three separate data channels that must be set: green, red, and blue. The green is used for the higher energy isotope, the red for the lower energy isotope, and the blue for the isotope ratio. In the double-isotope mode, high E^2/B is maintained by lowering the upper window of the higher energy isotope automatically to include the spectrum of that isotope and also to exclude the additional background counts, which may otherwise be quenched into the window if the discriminator window width is unchanged. As with any external standard ratio method for quench correction, the conditions determined for automatic analysis apply only to experimental samples using the same solvent system and counting vials used to set the ten points of the quench curve.

When a sample is actually being counted with this system, the following events take place. First, the sample is loaded into the machine. Then the external standard pellet is exposed and the external standard channels ratio is determined. Let us assume a ratio of 0.7325 is found. The automatic analyzer then knows that the level of quench desired is 0.7000 from the first digit of the external standard ratio. From the next three places, 0.0325, the analyzer knows that it must increase the magnetic field to reduce the external standard channels ratio by this amount. It does this in a series of 6-second trials and, when the external standard ratio of 0.7000 ± 0.0040 is obtained, a 30-second final count is made to provide the final precise electronic adjustment. The sample is then counted for the prescribed time or count set by the investigator and, when the counting is finished, both the raw and computed data are printed.

For samples of uniform quench characteristics, a rapid computational mode is available wherein the computed net counts per minute are converted electronically to disintegrations per minute by simple multiplication by the appropriate efficiency factor without the intervention of the normal quench adjustment routine. In other words, the net counts are simply multiplied by a constant and the results as disintegrations per minute are printed.

Automatic Quench Correction (AQC)

Another approach to automated counting efficiency determination, introduced by Beckman, is based on the premise that more accurate determinations of counting efficiency can be made if (a) the pulse height distribution of the counts altered by quenching can be restored to the unquenched level and (b) this pulse height restoration is essentially a linear function between the most-quenched to the least-quenched samples (9). The advantage of this approach is especially evident when double labeled samples are considered. As quenching becomes severe, for example in a mixture containing tritium and carbon-14, the carbon-14 spectrum can be completely quenched out of its counting channel, making the analysis impossible because all of the tritium and carbon-14 counted are in the tritium counting channel. With AQC, the spectrum can be amplified automatically so that a reasonable amount of the carbon-14 spectrum is restored into its counting channel, thereby facilitating analysis of the sample.

In practice, the AQC system is set up as follows: The least-quenched sample to be analyzed is first counted using the external standard. The external standard ratio for this least-quenched sample is then stored into a register. Next, the most-quenched sample to be analyzed is counted. Again, the external standard ratio for this most-quenched sample is manually stored into a second register. The difference between the two ratios is converted into an analog signal that sets the slope of the gain restoration function when the increased gain on the photomultiplier tube necessary to restore the most-quenched spectrum is determined. The AQC system can now correct any quenched sample that is within the AQC range by generating an output from the photomultiplier tube, which is always equivalent to that of the least-quenched standard. Again, this entire system is based on the premise that the quench correction by AQC for any given range of quench is a near-linear function. This should be carefully checked if the range of quench being normalized is large.

Intertechnique ABAC SL40

This liquid scintillation counter is the first automatic liquid scintillation counter to be sold with a small built-in computer capable of doing quench correction by external standard channels ratio. The storage capacity of the computer of 1024 12-bit words allows the storage of four separate quench curves. A second version of the SL40 has an expanded computer memory storage capacity of 4000 12-bit words.

REFERENCES

1. Jarrett, A. A. "Statistical Methods Used in the Measurement of Radioactivity with some Useful Graphs and Nomographs," AECU-262. U.S. At. Energy Comm., Oak Ridge, Tennessee, 1946.
2. Bush, E. T. *Anal. Chem.* **35,** 1024 (1963).
3. Bush, E. T. *Anal. Chem.* **36,** 1082 (1964).
4. van der Laarse, J. D. *Int. J. Appl. Radiat. Isotop.* **18,** 485 (1967).
5. Turner, J. C. *Int. J. Appl. Radiat. Isotop.* **19,** 557 (1968).
6. Bush, E. T. *Int. J. Appl. Radiat. Isotop.* **19,** 447 (1968).
7. Houx, N. W. H. *Anal. Biochem.* **30,** 302 (1969).
8. Natrella, M. G. "Experimental Statistics," Nat. Bur. Stds. Handbook 91. U.S. GPO, Washington, D.C., 1963.
9. Wang, C. H. *In* "The Current Status of Liquid Scintillation Counting" (Bransome, E. D., Jr., ed.), p. 305. Grune & Stratton, New York, 1970.

5

Basic Principles of the Application of Radiotracers

It is unusual nowadays for a scientist to synthesize his own labeled compounds. Commercial suppliers have established reasonable standards and are better equipped to produce radiochemicals of quality more economically. However, whatever the source of the radiochemicals, it is still prudent to establish the purity and stability of the compounds before commencing an extensive research project dependent on them (1–3). Further purification may be routinely necessary in some cases. For example, an enzyme substrate with a 1% impurity presents severe difficulties if the conversion rate is of the order of 0.1% and the impurity and the end product are not easily separated.

The greatest source of impurities is the decomposition of the labeled compounds during storage. This may be caused by self-radiolysis and/or free-radical reactions (4). This is especially true for tritiated compounds with high specific activities.

When a tracer is introduced into a biological system, it is assumed that

a. The tracer is indistinguishable from the unlabeled counterpart and therefore does not influence normal metabolism.

b. The labeled atom of the molecule remains in its original position; i.e., it does not "wander" from one atom to another or exchange with the solvent.

c. The tracer is sufficiently active to withstand dilution in the biological system so that changes can be accurately determined.

d. Methods are available for the isolation, purification, identification, and quantitation of the metabolites under study.

ISOTOPE DILUTION

Almost all radioassay methods employ a form of the isotope dilution principle, which transforms purification procedures into quantitative methods of analysis. In its simplest form, the tracer is used to monitor recovery (or measure the losses) sustained during the isolation of a particular compound in pure form. For example, a determination of the amount of A in a mixture containing A, B, C, D, etc., can be made provided that procedures are known for the isolation and purification of A and that it is available in radioactive form that we designate A*.

Example

Step 1: Add 10,000 dpm of A* (1 μg) to the mixture containing A, B, C, D, etc.

Step 2: Isolate and purify pure A + A* (i.e., by extraction, chromatography, crystallization, etc.)

Step 3: Determine the amount of pure A + A* (by weighing, fluorometry, spectrophotometry, or some appropriate means): 5 mg of A + A* found.

Step 4: Determine the radioactive content of A + A*; 2500 dpm recovered.

Calculations

The amount of A in original mixture is

$$\frac{(5 \text{ mg})(10{,}000 \text{ dpm})}{2500 \text{ dpm}} = 20 \text{ mg A}$$

Note that although 75% of the original amount of A in the mixture has been lost during the purification procedure, an accurate determination of the original content of A is possible. Also, another point is that the amount of tracer A* added to the mixture should represent only a small percentage of the total being measured (in this case 1 μg compared with a total of 20 mg so that it does not introduce any significant error in the final determination of purified A + A*). The critical aspects of this dilution procedure are that (a) A + A* be isolated in pure chemical form, (b) the weight of purified A + A* be accurately determined, and (c) the determination of radioactivity be within acceptable statistical limits.

Dilution analysis is based on the fundamental definition

$$\text{Specific activity} = \frac{\text{amount of radioactivity}}{\text{weight of material}}$$

In the example given, when 10,000 dpm of A* was added to the mixture, the specific activity of A* was changed by dilution with unlabeled A to

$$\text{Specific activity} = \frac{10{,}000 \text{ dpm}}{\text{wt of A in mixture} + \text{wt of A* added}}$$

Once the tracer is added and the sample thoroughly mixed, the specific activity of the tracer A* is diluted by the nonradioactive A (x mg) in the sample to a lower and uniform specific activity. Because the specific activity is a ratio, the requirements for its determination are the isolation of a measurable amount of pure A (which is now radioactive) and the determination of its weight and its radioactive content. Once the specific activity is determined on a purified sample of A, the original content of A in the sample is obtained by dividing the original radioactivity added to the sample (10,000 dpm) by the new specific activity.

In our example, 5 mg of pure A* was isolated, containing 2500 dpm. Therefore, its specific activity was

$$\text{Specific activity A*} = \frac{2500 \text{ dpm}}{5 \text{ mg A} + (5/X)(0.001 \text{ mg A*})}$$

where X = weight of A in the original sample.

However, because the fraction of the tracer $(5/X)$ (0.001 mg) contained in the 5-mg sample isolated is very small (less than 0.01%), it can be ignored. Then

$$\text{Specific activity of A} = \frac{2500 \text{ dpm}}{5 \text{ mg}} = 500 \text{ dpm/mg}$$

The original content of A (x mg) in the mixture can be calculated from the fact that the sample contains 10,000 dpm of A + A*, with a specific activity of 500 dpm/mg.

$$\text{Specific activity of A} = \frac{10{,}000 \text{ dpm}}{\text{wt } X} = 500 \text{ dpm/mg}$$

$$\text{Weight}_x = \frac{10{,}000 \text{ dpm}}{500 \text{ dpm/mg}} = 20 \text{ mg}$$

The equation for a simple dilution analysis can therefore be written

$$\text{Weight}_x = \frac{\text{dpm}_{\text{orig}}}{\text{sp act}_{\text{found}}}$$

or

$$\text{Weight}_x = \left(\frac{\text{dpm}_{\text{orig}}}{\text{dpm}_{\text{found}}}\right)(\text{wt}_{\text{found}})$$

In some instances, the weight of the tracer cannot be ignored because it makes a measurable contribution to the mass of the material being isolated. This may arise in a hypothetical situation in which an amino acid, B, is to be determined in 10 g of a protein and labeled B* is available only with low specific activity. In such a case, the content of the amino acid B in the protein can be determined after hydrolysis and dilution with B* if three parameters are known: (a) the weight of B* added, (b) the specific activity of undiluted B*, and (c) the specific activity of B* after dilution with B in the sample, which is derived from the following relationships.

$$\text{Specific activity}_{\text{final}} = \frac{\text{amount of radioactivity added}}{\text{wt}_{\text{B* added}} + \text{wt}_{\text{B,protein}}}$$

because

$$\text{Specific activity}_{\text{B*}} = \frac{\text{amount of radioactivity added}}{\text{wt}_{\text{B* added}}}$$

substituting for "amount" of radioactivity added

$$\text{Specific activity}_{\text{final}} = \frac{(\text{specific activity}_{\text{B*}})(\text{wt}_{\text{B* added}})}{\text{wt}_{\text{B* added}} + \text{wt}_{\text{B,protein}}}$$

then

$$\text{wt}_{\text{B,protein}} = \text{wt}_{\text{B*}}\left(\frac{\text{specific activity}_{\text{B*}}}{\text{specific activity}_{\text{final}}} - 1\right)$$

Example

Step 1: 10 mg B* containing 100,000 dpm are added to 10 g of protein containing B and the protein is hydrolyzed.

Step 2: B + B* isolated in pure form from hydrolyzate.

Step 3: Determine amount of B + B*: 80 mg recovered

Step 4: Determine radioactive content of B + B*; 12,800 dpm found.

Calculations

$$\text{Specific activity}_{B^*} = \left(\frac{100{,}000\ \text{dpm}}{10\ \text{mg}}\right) = 10{,}000\ \text{dpm/mg}$$

$$\text{specific activity}_{\text{final}} = \left(\frac{12{,}800\ \text{dpm}}{80\ \text{mg}}\right) = 160\ \text{dpm/mg}$$

$$\text{wt}_{B,\text{protein}} = \text{wt}_{B^*}\left(\frac{\text{specific activity } B^*}{\text{specific activity}_{\text{final}}} - 1\right)$$

$$= 10\ \text{mg}\left(\frac{10{,}000\ \text{dpm}}{160\ \text{dpm}} - 1\right)$$

$$= 615\ \text{mg of B in original mixture}$$

Therefore, the protein contained

$$\left(\frac{615\ \text{mg}}{10{,}000\ \text{mg}}\right)(100) = 6.15\%\ \text{B}$$

DERIVATIVE ANALYSIS USING A RADIOACTIVE REAGENT

A common problem in quantitating natural products, such as hormones and metabolites of biogenic amines, is one of concentration. The usual methods of isolation and identification are not applicable because the concentrations of these substances are beyond the limits of wet chemical methods or spectroscopic techniques. These substances can often be quantitated by reacting them with radioactive reagents to form labeled derivatives, which may then be characterized and analyzed by suitable analytical methods and the principles of isotope dilution.

Determination of the Radioactive Derivative without the Addition of Carrier

Sometimes, it is possible to isolate and determine the amount of a substance in a mixture without the addition of any carrier. In this instance, the substance can be quantitated on the basis of the specific activity of the reagent.

Example

Substance A reacts quantitatively with reagent R*, which is radioactive, forming a derivative R*A. The specific activity of R* is 100,000 dpm/μmole.

1. A mixture containing A is reacted with an excess of reagent R* (200,000 dpm).
2. Excess reagent R* is removed from the reaction mixture.
3. Reaction product R*A is isolated and purified and its radioactive content is measured; 50,000 dpm found.
4. How much A was in the original mixture?

Calculations

Because the specific activity of R* is 100,000 dpm/μmole and R* and A react on a mole per mole ratio, the amount of A in the mixture is

$$\left(\frac{50{,}000 \text{ dpm}}{100{,}000 \text{ dpm}/\mu\text{mole}}\right) = 0.5\ \mu\text{mole}$$

This result assumes (a) a quantitative reaction between R* and A and (b) a quantitative recovery of the product R*A.

Determination of the Radioactive Derivative with Addition of Carrier

In most isolation and purification procedures, it is rare to achieve quantitative recovery of the product. Such factors as adsorption of material on the reaction vessel walls, incomplete reaction between the reagent and the compound, and incomplete transfers of the reaction mixture between steps all contribute to the losses observed. These losses can be monitored by diluting the reaction product R*A with nonradioactive carrier RA. The only assumption is that the reaction between R* and A is quantitative.

Example

1. A mixture containing A is reacted with an excess of reagent R*, specific activity = 100,000 dpm/μmole.
2. After the reaction, 100 mg of carrier RA is added to the reaction mixture.
3. Excess reagent R* is removed from the reaction mixture.
4. Reaction product R*A + RA is isolated, purified, and weighed; 75 mg is recovered. The radioactive content of the total sample is found to be 50,000 dpm.
5. How much A was in the original mixture?

Calculations

Because 75 mg of the original 100 mg of carrier RA was recovered, 25 mg or 25% of the original RA was lost during the isolation and purification procedures. The 50,000 dpm recovered represents only 75% of the original and must be corrected back to 100%. This is done as follows.

$$(50{,}000 \text{ dpm})\left(\frac{100 \text{ mg}}{75 \text{ mg}}\right) = 66{,}666 \text{ dpm in original sample}$$

$$\left(\frac{66{,}666 \text{ dpm}}{100{,}000 \text{ dpm}/\mu\text{mole}}\right) = 0.66\ \mu\text{mole of A in original sample}$$

DOUBLE-ISOTOPE DERIVATIVE ANALYSIS

With the exception of the derivative methods, all the other isotopic methods so far described required some means of assessing the weight of the end product. The most powerful variation of the isotope dilution technique was the double-isotope derivative analysis. Here, the requirement for weighing the product was eliminated by the use of two isotopes: one to measure the amount of product and the second to monitor the isolation losses. This one technique was probably responsible for much of the advances made in steroid biochemistry during the last 15 years (5). Double-isotope methods, however, were tedious and required great care in their execution. When radioimmunoassay and binding protein methods were introduced, they were quickly adopted because they appeared simpler, less demanding, and more amenable to large sample populations. These methods will be discussed below.

The sensitivity of the double-isotope derivative method depends on the specific activity of the indicator or recovery compound and that of the radioactive reagent. The highest specific activity possible is with tritium, which can be obtained at 100% purity, or 58,276 Ci/mole. Carbon-14, in contrast, can be obtained only at about 60 Ci/mole. Its long half-life makes the production of pure carbon-14 unlikely. To attain the required sensitivity, the isotope used for the indicator is usually tritium and that for the reagent is carbon-14 or sulfur-35. Caution is always required when any material of high specific activity is used because the deterioration of the material by self-radiolysis is maximal under these conditions. Stability and purity of both the indicator and the reagent are essential for the success of this technique.

Example

In a mixture of A + B + C + D + · · ·, A is to be determined. An indicator, tritiated A, A*, is available at a specific activity of 40 Ci/mmole. Reagent R*, labeled with carbon-14, reacts with A and has a specific activity of 10,000 dpm/nmole.

1. Add 5000 dpm of tritiated A* (0.057 pmole) to the mixture as indicator.
2. React mixture with excess reagent R*.
3. Remove excess reagent R*.
4. Purify product, R*A* + R*A. Purity is indicated by a constant ratio of ^{14}C to ^{3}H.
5. Find ^{3}H counts (= 2000 dpm) and ^{14}C counts (= 500 dpm).
6. How much A was in the original mixture?

The tritium count of 2000 dpm shows that 5000 dpm less 2000 dpm or three fifths of the derivative has been lost during the purification because of, e.g., incomplete reaction between R* and A + A* and isolation losses. The final carbon-14 counts of 500 dpm must be corrected to what the counts should be were the reaction quantitative and the recovery 100%. The corrected carbon-14 count is

$$(500 \text{ dpm } ^{14}\text{C})\left(\frac{5000 \text{ dpm } ^{3}\text{H}}{2000 \text{ dpm } ^{3}\text{H}}\right) = 1250 \text{ dpm } ^{14}\text{C}$$

The 1250-dpm ^{14}C is stoichiometrically equivalent to the total amount of A in the original sample.

$$\left(\frac{1250 \text{ dpm } ^{14}\text{C}}{10{,}000 \text{ dpm } ^{14}\text{C/nmole}}\right) = 0.125 \text{ nmoles or } 125 \text{ pmoles of A}$$

In this example, it is not necessary to correct for the carbon-14 that reacted with the radioactive indicator A* because this correction is negligible. The 2000 dpm of tritium in the recovered final sample is equivalent to (2000/5000) (0.057 pmoles) or 0.038 pmoles of A*, whereas the amount of unlabeled A in the same sample is equivalent to (500/10,000) (1000 pmoles) or 50 pmoles.

In this type of double-isotope analysis, the reaction between the reagent R* and the compound A need not be quantitative. The major requirement is that the product R*A be properly purified. Purity is usually established by demonstrating the achievement of a constant $^{3}H/^{14}C$ ratio or by using a rigorous purification scheme proved to result in the recovery of a single product.

The sensitivity of double-isotope methods are usually in the nano- and picogram range because of the very high specific activity indicators now available.

RADIOCHEMICAL SATURATION ANALYSIS

Another variation of the isotope dilution analysis is saturation analysis. The basic outline of the method is as follows: A known weight of A (the compound to be analyzed) in radioactive form, A*, is added to the sample. A* is the indicator for the procedure. A known quantity of a reagent R that specifically and quantitatively reacts with A is then added to the mixture in less than the stoichiometric amount of A expected in the mixture, so that all of the reagent R is saturated with compound A + A* to form the product, RA + RA*, leaving an excess of A + A* in the mixture unreacted. The product RA + RA* is then separated and analyzed for its radioactive content. The specific activity of the unknown product, S_{unknown}, is then equal to the counts observed in the product divided by the weight of the product RA + RA*. The procedure is now repeated but this time only an excess of the undiluted indicator A* is reacted with the same amount of the reagent R used above, and the product RA* is isolated as before. The specific activity of this product, S_{standard}, becomes the reference standard. As the weight of the product isolated in this procedure is always the same because a constant amount of the reagent R is used, the ratio of the two specific activities can be related directly to their respective counts.

$$\frac{S_{\text{standard}}}{S_{\text{unknown}}} = \frac{\text{cpm}_{\text{standard}}}{\text{cpm}_{\text{unknown}}}$$

This identity can then be substituted directly into the fundamental isotope dilution equation derived previously to give the expression

$$W_{\text{unknown}} = W_{\text{standard}}\left(\frac{\text{cpm}_{\text{standard}}}{\text{cpm}_{\text{unknown}}} - 1\right)$$

The weight of the standard, W_{standard}, of course, is derived from the knowledge of the specific activity of the indicator A* and the quantity of radioactivity used as the indicator in the procedure.

The critical assumptions of this method are that (a) the radioactive indicator A* and the compound A are indistinguishable, (b) the total amount of compound A + A* is always in excess, stoichiometrically, to the reagent R, (c) the amount of product RA + RA* isolated is always the same, and (d) the product RA + RA* is not contaminated with unreacted A*.

SATURATION ANALYSIS APPLIED TO BIOLOGICAL SYSTEMS

An important advance in the quantitation of biological compounds in the past decade has been the introduction of a group of analytical methods that exhibit marked improvements in both sensitivity and specificity. In these procedures the sensitivity of radioactive determinations is combined with a chemical specificity that serves to characterize the substance involved. These methods have come to be known by a variety of different names, including "radioimmunoassay" (6, 7), "saturation analysis" (8, 9), "competitive protein-binding methods" (10, 11), "radioenzymatic isotope dilution assays" (12–14), and "displacement analysis" (15). None of these terms is entirely satisfactory. The binding agent, for example, need not be protein or immunologically induced. Ekins (9) has advocated the term "saturation analysis" because, in general, these methods depend on the progressive saturation of a specific binding reagent.

The fundamental approach has been developed independently in two laboratories. It has been applied, on the one hand, by Yalow and Berson (6) to the assay of insulin using a specific antibody and on the other, by Ekins (8) to the assay of plasma thyroxine, using a naturally occurring binding protein. The principle is the same in both cases. When a mixture of radioactive and nonradioactive form of the same chemical substance is introduced into a system containing a specific binding agent, such as an antibody, then some of the compound reacts with the binding agent and some of it remains free. If the concentrations of labeled indicator and the amount of binding agent are fixed, then the higher the concentration of the unlabeled substance, the less the amount of labeled substance bound to the protein. Therefore, when the radioactivity of the protein-bound substance is measured with and without the addition of known amounts of nonradioactive material a standard curve can be constructed. Unknowns are estimated by comparison with the standard curve.

The major variations of saturation analysis are

1. Radioimmunoassays
2. Competitive protein binding methods
3. Radioenzymatic assays

In the simplest case the relationship between the binding reagent R and the compound A is governed by the law of mass action

$$A + R \rightleftharpoons AR$$

At equilibrium

$$\frac{[AR]}{[A][R]} = \frac{k_1}{k_2} = K$$

where K is the equilibrium constant. If binding is great, K is large, indicating a high affinity of the protein for the ligand.

This type of analysis requires

1. The compound A to be available in radioactive form A*
2. A specific binding reagent, R
3. A procedure for sequestering free from bound moieties

The original radioimmunoassay described by Yalow and Berson (6) depends on the competitive inhibition of binding between the antibody and the specific radioactive antigen by unlabeled antigen. This technique has now been applied to a variety of polypeptide hormones. For compounds with a molecular weight of less than 3000, it has usually been necessary to couple the peptide to a larger molecule, such as a protein or polylysine, in order to induce antibodies.

A typical antibody dilution curve is shown in Fig. 54. As the protein is diluted the percentage of labeled indicator bound to the protein decreases. The concentration of antibody used for assay is usually that which binds 50% of the labeled indicator.

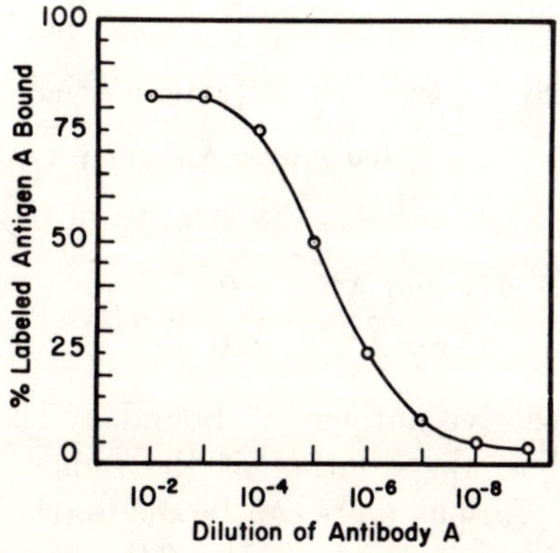

Fig. 54 A typical antibody dilution curve for a hypothetical antibody A.

The Standard Curve

Let us assume that the antibody is ideal and does not exhibit any nonspecific binding. We further stipulate that the amount of labeled antigen A* used is 200 units per tube (the same amount used in the antibody dilution curve) and that, following equilibration, a method is available to quantitatively separate the free labeled antigen A* from that bound to the specific antibody. At equilibrium, when no unlabeled antigen A is added, the following relationships exist.

$$\text{200 antigen A* (free)} + \text{antibody A} \rightleftharpoons \text{100 [antigen A*–antibody A complex]} + \text{100 antigen A* (free)}$$

The ratio of the free antigen A* to bound antigen A* at equilibrium is unity. The 100 units of labeled antigen A* bound expressed as counts per unit time also represent the zero point value, i.e., where no unlabeled antigen A has been added. To set up a standard curve, a series of assay tubes are prepared in duplicate or triplicate, each containing 200 units of radioactive antigen A*, a constant amount of antibody A, and varying amounts of unlabeled antigen A. The tubes are then allowed to equilibrate at 4°C. After equilibration, fractions containing the bound and free antigen A* are separated from one another and the radioactive content of the appropriate fraction determined. For example, if 200 units of unlabeled antigen A are added to an assay tube, there is a total of 400 units of antigen A (200 labeled and 200 unlabeled). Because the amount of antibody A is constant, the antibody binds 100 units of antigen A, but of these 100 units bound only 50 are labeled.

$$\text{200 antigen A} + \text{200 antigen A*} + \text{antibody A} \rightleftharpoons \text{50 [antigen A–antibody A complex]} + \text{50 [antigen A*–antibody A complex]} + \text{150 antigen A (free)} + \text{150 antigen A* (free)}$$

Therefore, the ratio

$$\frac{\text{Bound A*}}{\text{Free A*}} = \frac{50}{150} = 0.33$$

The amount of radioactive antigen A* bound by the antibody, then, is directly proportional to the dilution of the radioactive antigen A* by unlabeled antigen A. Various plots can be made relating the distribution of the radioactivity between the bound and the free state to the concentration of unlabeled antigen A added to the system. A theoretical table of values based on the conditions described here is given in Table 18. The

TABLE 18

Theoretical Values Based on Conditions Described in the Text

Unlabeled antigen A	Labeled antigen A*	Free (F) antigen A*	Percent free A*	Bound (B) antigen A*	Percent bound A*	Ratio (B/F)	Ratio (F/B)	Logit Y^a
0	200	100	50	100	50	1.00	1.00	—
10	200	105	52.5	95	47.5	0.90	1.11	2.9444
20	200	109	54.5	91	45.5	0.83	1.20	2.3136
30	200	113	56.5	87	43.5	0.76	1.30	1.9010
40	200	117	58.5	83	41.5	0.70	1.41	1.5856
50	200	120	60	80	40	0.66	1.50	1.3863
60	200	123	61.5	77	38.5	0.62	1.60	1.2083
70	200	126	63	74	37	0.58	1.70	1.0460
80	200	129	64.5	71	35.5	0.55	1.82	0.8954
90	200	131	65.5	69	34.5	0.52	1.90	0.8001
100	200	133	66.5	67	33.5	0.50	1.99	0.7082
200	200	150	75	50	25	0.33	3.00	0
300	200	160	80	40	20	0.25	4.00	−0.4055
400	200	167	83.5	33	16.5	0.19	5.06	−0.7082
600	200	175	87.5	25	12.5	0.14	7.00	−1.0986
1000	200	183	91.5	17	8.5	0.09	10.76	−1.5856
2000	200	191	95.5	9	4.5	0.05	21.22	−2.3136

[a] Logit $Y = \ln[Y/(100 - Y)]$, where Y = percent of zero radioactive antigen A* bound value.

most widely used plots for constructing a standard curve from these data are shown in Fig. 55A–E. They are (A) the percent radioactive antigen A* bound versus the concentration of unlabeled antigen A added, (B) the ratio of the radioactive antigen A* bound to the free radioactive antigen A* versus the concentration of unlabeled antigen A added, (C) the ratio of radioactive antigen A* free to bound versus the concentration of unlabeled antigen A added, and (D) the percent of zero radioactive antigen A* bound value versus log concentration of unlabeled antigen A added. The percent of zero radioactive antigen A* bound value is computed by taking the amount of radioactive antigen A* bound in the assay system when no unlabeled antigen A is added as 100%. Figure 55E is the logit Y versus log concentration of unlabeled antigen A. Logit Y is defined as follows.

$$\text{logit } Y = \log_e \frac{[Y]}{100 - Y}$$

where Y is the percent of zero radioactive antigen A bound value as defined in (D) above. Inspection of the various standard curves shown in Fig. 55 shows that curves C and E are linear. Curve C, the plot of the ratio of the bound to the free antigen A*, implies that the radioactivity in both fractions are measured.

The measurement of both bound and free moieties reduces the possibility of inadvertent error arising from the introduction of unequal amounts of radioactive tracer into all incubation tubes. The pressure of counting time, however, is such that usually only one fraction is counted. From the counting standpoint, the fraction with the greater activity is counted with the greatest precision but the change in activity for a given change in hormone concentration are greatest in fractions containing the lesser activity. It is therefore preferable to count samples containing the least activity for best accuracy in determining the response to a change in antigen concentration.

The logit transform, curve E, is a frequently used transformation in this kind of analysis because of its linear nature. The computed logit values are fitted to a simple regression line. However, in using a logit plot, one must be aware that the lower end of the curve has far less sensitivity than the upper end of the curve. For example, in Fig. 55E, a change in the logit value at the upper end from 2.5 to 2.0 is equivalent to a net change of 10 units of antigen A, whereas a logit change from -2.0 to -2.5 on the lower end is equivalent to a change of 1000 units of antigen A.

A plot of the percent radioactive antigen bound versus the log of the concentration of the unlabeled antigen is shown in curve D, Fig. 55. The linear portion of the curve lies between 70 and 400 units. The points

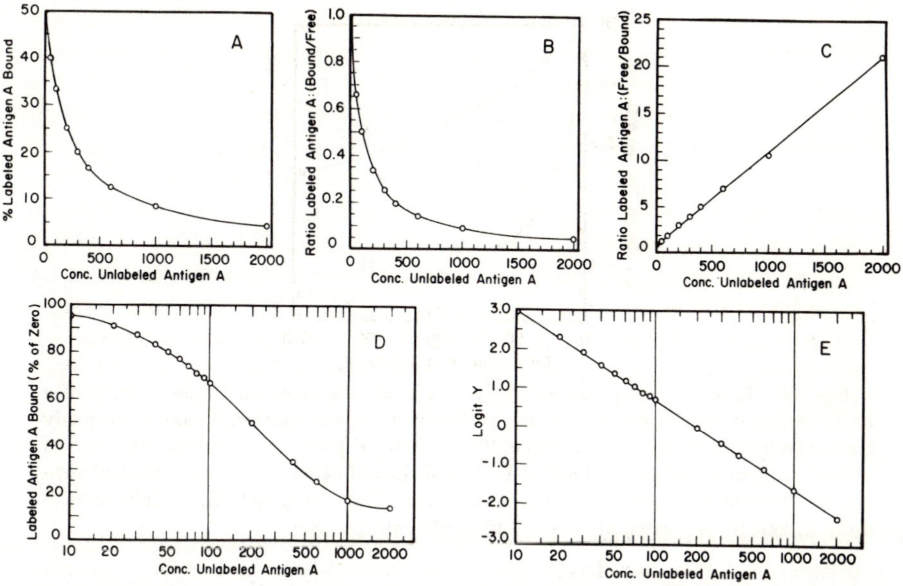

Fig. 55 Various standard curves for a hypothetical radioimmunoassay of an antigen A. (A) Percent radioactive antigen A* bound versus concentration of unlabeled antigen A. (B) Ratio of radioactive antigen A* bound to free versus concentration of unlabeled antigen A. (C) Ratio of radioactive antigen A* free to bound versus concentration of unlabeled antigen A. (D) Percent of zero radioactive antigen A* bound value versus log concentration of unlabeled antigen A. (E) The logit *Y* versus log concentration of unlabeled antigen A.

above and below these units are not considered useable because they deviate from the straight line. However, it can be seen that this plot is, in fact, part of a nonlinear, S-shaped curve. The points describing the entire curve are identical to those used in the logit transform in curve E.

A standard curve based on a plot of the counts in the bound fraction versus the concentration of the unlabeled antigen is equivalent in shape to curve A, Fig. 55. Mathematically, the curve can be described by an equation for a hyperbole (16).

Alteration of the Standard Curve

Increase in sensitivity of a particular assay can be achieved in several ways: (a) by decreasing both the amount of antibody and labeled antigen used, (b) by using a larger sample, or (c) by increasing the specific activity

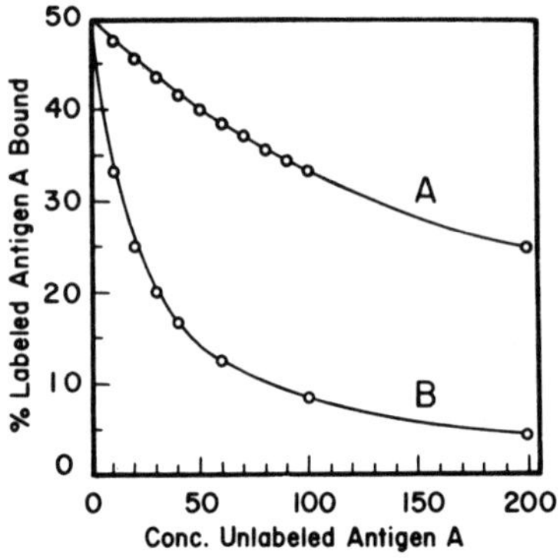

Fig. 56 Effect of using less antibody and labeled antigen on the sensitivity of a hypothetical immunoassay system for antibody A. In this system, 1 unit of antibody binds specifically to 1 unit of antigen. The system in curve (A) contains 200 units of labeled antigen A, 100 units of antibody A, and the indicated amount of unlabeled antigen A. The system in curve (B) contains 20 units of labeled antigen A, 10 units of antibody A, and the indicated amounts of unlabeled antigen A.

of the antigen. The limiting factor in determining the sensitivity of a particular radioimmunoassay is usually the specific activity of the antigen available.

The result of increasing sensitivity by decreasing the amount of both antibody and labeled antigen is illustrated graphically in Fig. 56. The hypothetical amounts used are one tenth of those given in Table 18 and are listed in Table 19 for a range of 0–200 units of unlabeled antigen A. The calculations are based on the assumption that 50% of the labeled antigen A is bound by the antibody when no unlabeled antigen A is added. It is evident that with the diluted system, higher sensitivity is achieved as measured by the greater change in the percent of radioactive antigen A* bound per unit change of added unlabeled antigen A to this system. The steeper the slope, the greater the sensitivity. A family of such curves can be drawn by altering the fixed amounts of antibody and labeled antigen used in any such assay procedure. A consequence of increased sensitivity is a proportionate reduction in the useful range of the assay.

The Labeled Antigen (Ligand)

The labeled antigen (ligand) is the most critical component of the assay system because it is the component that is actually analyzed by counting. As in any analytical technique, the labeled antigen should be stable and have the highest chemical purity, specific activity, and affinity for the binding agent.

TABLE 19

Hypothetical Values of Antibody and Antigen for Figure 56

Unlabeled antigen A	Labeled antigen A*	Free (F) antigen A*	Percent free A*	Bound (B) antigen A*	Percent bound A*	Ratio (B/F)	Ratio (F/B)	Logit Y^a
0	20	10	50	10	50	1	1	—
10	20	13.33	66.65	6.67	33.35	0.5003	1.9985	0.6946
20	20	15	75	5	25	0.3333	3	0
30	20	16	80	4	20	0.25	4	−0.4055
40	20	16.67	83.35	3.33	16.65	0.1997	5.006	−0.6946
60	20	17.5	87.5	2.5	12.50	0.1428	7	−1.0986
100	20	18.33	91.65	1.67	8.35	0.0911	10.976	−1.6070
200	20	19.09	95.45	0.91	4.55	0.0476	20.978	−2.3015

[a] Logit $Y = \ln[Y/(100 - Y)]$.

The choice of isotope has usually been from the following: ^{131}I, ^{125}I, ^{3}H, and ^{14}C. Based on the incorporation of an atom of radioisotope per molecule of antigen (ligand) the maximum specific activities in curies per millimole of antigen that are possible are shown in Table 20.

The label of choice has been radioiodine because of the high specific activity available, the ease of iodinating almost any protein, and the simplicity of radioassay by gamma counting. Of the two primary radioiodines commonly used, ^{125}I is preferred over ^{131}I because of the longer half-life available, although both isotopes have their own attributes. For example, although ^{125}I is a weaker gamma emitter than ^{131}I, it can be counted at high counting efficiencies under certain conditions, the lower gamma energies favor less autodecomposition of the labeled antigen, and it is a much safer radioisotope to handle. Ability to store the labeled antigen is a desirable feature of ^{125}I but it is balanced by damage caused by autoradiolysis during storage. With ^{131}I, higher specific activity is possible and labeling is done for each analysis because of its relatively short half-life. This latter procedure can conserve a valuable antigen, which can be stored unlabeled indefinitely and labeled as required. The choice of the radioisotope is a balance among the following considerations: (a) availability of the antigen; (b) the time required for iodination and isolation of the labeled antigen; (c) the specific activity of the product, which determines the counting time required; and (d) the cost of the isotope. ^{131}I costs about half as much as ^{125}I.

TABLE 20

Radionuclides Used in Protein-Binding Assays

Radionuclide	Half-life	Maximum specific activity (Ci/mmole)	Gamma energies (MeV, %)	Beta energies (MeV, %)	Decay product
Tritium	12.3 years	29.4	—	0.018 (100)	^{3}He
Carbon-14	5.73×10^3 years	0.064	—	0.155 (100)	^{14}N
Phosphorus-32	14.3 days	9,760	—	1.70 (100)	^{32}S
Phosphorus-33	25 days	5,280	—	0.25 (100)	^{32}S
Sulfur-35	86.7 days	1,400	—	0.17 (100)	^{35}Cl
Iodine-125	60.2 days	2,125	0.27 X rays 0.035 (7)	$0.03e^-$ (90)	^{125}Te
Iodine-131	8.05 days	15,720	0.36 (80) others to 0.72	0.61 (87) others to 0.81	^{131}Xe

So far, the assumption has been that the labeled and unlabeled hormones are identical in their reaction energy with the binding site. This is usually true for tritiated indicators, in which the substitution of the isotope has little or no effect on the configuration of the molecule. With radioiodine, however, the iodination of the tyrosine residues in a polypeptide, for example, may produce a labeled indicator that differs in its reaction energy from the unlabeled hormone. This difference in avidity does not invalidate a saturation assay but it may affect the optimal concentration of reactants required to achieve maximum sensitivity.

Tritium-labeled antigens have been used mainly for the steroid hormones. Although the specific activity of any tritiated material is considerably less than that for iodinated antigens, it is sufficiently high to serve as a useful indicator in some cases. The storage life of tritiated antigens is relatively long and, consequently, it is not necessary to standardize a tritiated antigen every time an assay is run.

Carbon-14 labeled antigens are not used because of their low specific activity. Other potentially useful radionuclides for protein-binding assays are ^{32}P, ^{33}P, and ^{35}S. Their nonradioactive counterparts are constituents of important biological materials and the radionuclides are available with specific activities well suited for this application.

Separation of Bound from Free Antigen

The critical step in all protein-binding methods is the separation of free from bound antigen (ligand). The precision and reproducibility of the method depends on the establishment of equilibrium between the bound and free antigen and on the maintenance of this equilibrium during the separation procedure. The procedure usually involves the addition or use of a third agent to facilitate the quantitative separation of the two fractions. These agents include (a) a second antibody to selectively precipitate the antigen–antibody complex; (b) adsorbents, such as powdered talc, magnesium silicate, charcoal coated with dextran, or ion-exchange resins, to adsorb all the free antigen, which can then be removed by centrifugation; and (c) a solid support to adsorb the antibody. Solid-phase systems include antibodies fixed or bonded onto walls of glass or plastic tubes, separation using a cellulose ester filter, such as Millipore, or adsorption onto various types of glass or plastic beads. This technique has an obvious advantage in the ease of separating the free from the bound antigen. (d) Such chemical precipitants as ammonium sulfate, sodium sulfate, acetone, alcohol, or dioxane, are also needed to precipitate the antigen–

antibody complex. These agents are protein precipitants and are therefore not specific for any given antigen–antibody complex. All four of these methods are in current use.

EVALUATION OF A RADIOIMMUNOASSAY

The various criteria for precision, accuracy, sensitivity, and reproducibility as they apply to radioimmunoassay have been debated among the various theoreticians in the field. Yalow and Berson (17–19) have published an extensive mathematical analysis of the antigen–antibody reaction and have arrived at theoretical principles relating to optimized assay design for high sensitivity and precision. However, Ekins *et al.* (20–24) have claimed that these principles are fallacious. These differences center mainly on the definitions of assay sensitivity and precision. Some of these problems have been alluded to in the discussion of the various forms of the standard curve that can be used to quantitate unknowns by radioimmunoassay. Sensitivity has been defined by Midgley (25) as the smallest amount of unlabeled antigen that can be distinguished from no antigen. A mathematical theory of immunoradiometric assays has recently been published by Rodbard and Weiss (26).

Instrumentation for Radioimmunoassays

The majority of radioimmunoassays use either ^{125}I or ^{131}I, which are best counted in a gamma counter. For gamma emitters, there is no sample preparation required when using a gamma counter. However, gamma emitters can be counted efficiently in a liquid scintillation counter by using a toluene-based counting solution loaded with a heavy metal salt, such as those of tin or lead. This counting solution can then function in a manner similar to a crystal in a gamma counter. An encapsulated vial containing such a mixture for use in this application is commercially available but at a high cost. The sealed glass vial contains a central cavity for accepting a small plastic tube containing the sample. The design was first suggested by Ashcroft (27). Another experimental sample holder, consisting of a solid plastic scintillator loaded with lead salts with a hole drilled in the center to accept a plastic sample tube, has been described by Litt (28), who reports counting efficiencies in excess of 40% for ^{125}I. A third approach is the immersion of a small ^{125}I-containing sample tube (microfuge tube) directly into a 20-ml scintillation vial containing a toluene-based scintillator that includes 5% tetrabutyllead. An objection to this last approach is that when this vial is not holding a microfuge tube,

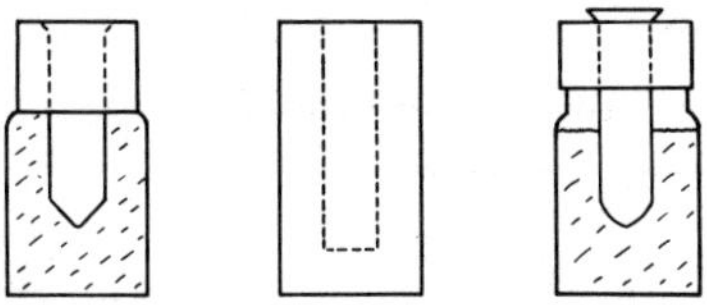

(A) (B) (C)

Fig. 57 Holders for counting ^{125}I-containing samples in a liquid scintillation counter. (A) An encapsulated vial with a central cavity for a sample tube. The vial is completely filled with a toluene-based scintillator containing a tin salt. (B) A solid plastic scintillator (Pilot Chemical Co.) loaded with 5% tetrabutyllead with a central cavity. (C) A 20-ml scintillation counting vial containing a toluene-based scintillator with 5% tetrabutyllead. The cap has a hole to accommodate a plastic microfuge tube.

the scintillator solution can concentrate by evaporation of the toluene through the hole in the cap. Another inconvenience is that every microfuge tube inserted into the vial is wetted with scintillator solution and is therefore difficult to handle. The three types of sample holders are shown in Fig. 57. The use of these special sample holders appears attractive but their uniformity of performance in terms of counting efficiency and reproducibility has not been established. Although inconvenient compared to direct gamma counting, ^{125}I-containing samples can be counted at efficiencies in excess of 50% using a normal counting solution used for accommodating aqueous samples.

COUNTING GAMMA EMITTERS BY LIQUID SCINTILLATION COUNTING

Although gamma emitters, such as iodine-131 and cobalt-60, are counted efficiently and conveniently in a gamma counter (which uses a sodium iodide crystal detector), these and other gamma emitters of clinical interest can be counted at equal or greater efficiency by liquid scintillation counting by taking advantage of the fact that these radioisotopes emit beta particles or Auger electrons, as well as gamma-ray photons, during their decay. A definitive example of this difference in counting efficiency of a gamma emitter by the two methods is illustrated by chromium-51, which decays by electron capture.

The distribution frequencies of the radiations emitted by chromium-51 are given in Table 21 (29). In a gamma counter, the theoretical maximum counting efficiency for chromium is the sum of the gamma- and X-ray disintegration frequencies (about 29%). The theoretical maximum for beta counting by liquid scintillation counting is the sum of the frequencies of occurrence of internal conversion, Compton electrons, and Auger elec-

TABLE 21

Radiations from Chromium-51 Decay[a]

Type of radiation	Energy (keV)	Disintegration frequency (%)
Gamma	323	9
Internal conversion electron	317.5	0.013
Vanadium X rays	5	20
Auger electrons	4.5	70

[a] Data taken from Sheppard and Marlow (29).

trons (about 100%). Electron capture by the nucleus of a K electron does not necessarily result in any nuclear radiation. The characteristic radiation of X rays emitted is the result of the rearrangement of the other orbital electrons to fill immediately the vacancy created by this process in the K shell. In addition, the emission of negative electrons, called "Auger electrons" after the French physicist P. Auger, can result in a situation in which the emission of X-ray photons is replaced by the emission of these electrons with a kinetic energy equal to the difference in binding energy of the electrons in the K and L shells. The internal conversion electrons result when a gamma photon fails to escape the electron cloud surrounding the nucleus and transfers sufficient energy to one of the orbital electrons to eject it from the atom. The Compton electrons are generated by the interaction of gamma photons with the counting solution and the vial. In practice, typical counting efficiencies by gamma counting using a NaI crystal are 15% and are approximately 70% by liquid scintillation counting. Gamma emitters that can be counted with high efficiency in a liquid scintillation counter using a normal counting solution (one not containing lead, tin, or other heavy metals) include iodine-131, iodine-125, cobalt-60, cobalt-57, iron-59, and chromium-51. It should be evident that, for liquid scintillation counting, those gamma emitters which also emit beta particles or Auger electrons should also be classed as beta emitters together with carbon-14 and tritium.

ENZYME–ISOTOPE DILUTION ANALYSIS

There have been relatively few examples of radioenzymatic assays that employ the principle of saturation analysis, whereby the enzyme substrate is in excess and the enzyme is rate limiting. This technique has wide application, although its potential has not been fully developed. Radioenzymatic assays have also been termed "enzyme isotope dilution analysis" and the theoretical principles have been enunciated independently by Newsholme and Taylor (30) for the assay of serum glycerol and by Brooker and Appleman (31, 32) for the determination of cyclic AMP using phosphodiesterase. The principle of enzymatic isotope dilution depends on (a) a decrease in product labeling according to the isotope dilution principle and (b) the relationship between the Michaelis–Menton constant, K_m, and the substrate concentration. If only a small fraction (10–20%) of substrate is converted to product, a simple mathematical relationship can be derived that relates the change in incorporation of the label into the product to the concentration of unlabeled substrate and the K_m of the enzyme. The procedure assumes a method for separating the unreacted substrate from the product of the reaction. The mathematical expression (30) for the method is

$$\frac{P_o}{P_s} = 1 + \frac{S}{K_m + S_o}$$

where P_o = total amount of radioactivity incorporated into product at substrate concentration S_o

P_s = total amount of radioactivity incorporated into product at substrate concentration $(S_o + S)$

S = unlabeled substrate concentration

S_o = labeled substrate concentration

K_m = Michaelis–Menton constant

The attraction of this method is that the specificity is provided by the enzyme and the sensitivity is a function of the K_m. The minimum concentration of S, in the above equation, that can be determined is about 10% of the K_m for the substrate. The method depends on the knowledge of the K_m of the substrate, especially if more than one substrate is involved, because the K_m of the substrate can vary according to the concentration of the second substrate.

The enzyme isotope dilution method is illustrated by the determination of endogenous L-histidine content in tissues using a bacterial enzyme, histidine decarboxylase, and carboxyl-labeled L-histidine. The product formed, radioactive carbon dioxide, is easily separated from the substrate and can be quantitatively recovered and counted.

Ten micrograms of carboxyl-labeled L-histidine (100,000 dpm) are incubated at 37°C with bacterial histidine decarboxylase and varying amounts

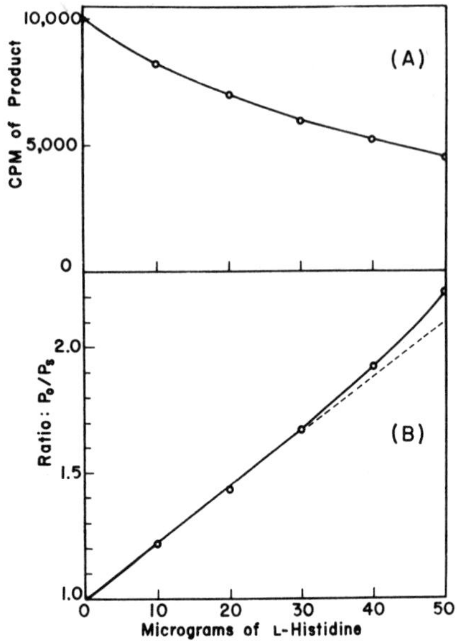

Fig. 58 An enzyme-isotope dilution curve for the determination of L-histidine. Curve A represents the amount of radioactive carbon dioxide evolved by a bacterial histidine decarboxylase versus the concentration of nonisotopic L-histidine added after 30 minutes incubation at 37°C in air. Curve B is a plot of the ratio P_o/P_s against the concentration of nonisotopic L-histidine. The dashed line shows the deviation from linearity after 30 μg of L-histidine.

of nonisotopic L-histidine, as indicated in Fig. 58. The reaction is stopped after 30 minutes and the radioactive carbon dioxide is liberated by acidification and trapped on a filter paper impregnated with Hyamine hydroxide (33). The results are shown graphically in Fig. 58A, where the counts recovered are plotted against the concentration of unlabeled L-histidine. In Fig. 58B, the ratio P_o/P_s is plotted against the concentration of unlabeled L-histidine. The straight line is similar to the plot of the free to bound ratio for radioimmunoassays.

RADIOENZYMATIC ASSAYS

Another area of rapid growth in recent years has been in the application of radioisotopes to the measurement of enzyme activities and, conversely, the use of enzymes for the quantitation of chemical compounds. Both Reed (34) and Oldham (35) have performed a great service by collating the information available on radiochemical methods of enzyme assay.

The enzyme and radioactive substrate are incubated for a known period of time together with appropriate blank samples. The reaction is stopped, often by acid precipitation of the enzyme, and the unreacted substrate and product are separated from one another. An aliquot of the isolated product is then taken for radioassay. In only a few cases have continuous assay methods been used in radioisotopic enzyme procedures (35).

The inherent sensitivity of these isotopic methods is especially attractive when a specific reaction is to be studied in detail. Furthermore, the formation of a radioactive product enables isotope dilution analysis to be applied to establishing the specificity of the reaction. A wide range of substrate and enzyme concentrations can be used and it is often possible to measure low enzymatic activity in crude homogenates without the need for extensive purification of the enzyme. This undoubtedly has constituted one of the major practical advantages of the radiochemical techniques.

The use of high specific activity substrates does not necessarily increase the sensitivity. If we define sensitivity in the same manner as for saturation analysis, i.e., the minimum amount detectable in the assay system, then for a known rate of reaction higher specific activity substrates may be expected to give greater amounts of radioactivity in the reaction product. This is usually the case but the increase may be offset by increased levels in the "blank" reading. High blank readings may be caused by impurities in the labeled substrate or by concomitant nonenzymatic reactions occurring during the incubation. The latter have been shown to be significant factors in assays of tyrosine aminotransferase (36), histidine decarboxylase (37, 38), and tyrosine hydroxylase (39). In addition, the problems of stability and purity of the labeled substrate becomes more acute as the specific activity is increased and greater demands are placed on the separative procedures used in isolating the end product. In discussing these problems, Oldham (40, 41) has recommended that labeled substrates be used at the lowest specific activity compatible with their proposed use.

A major undesirable feature of radiotracer assays is the time required to separate the unreacted substrate from the product formed. The separation procedures used include precipitation methods, chromatographic techniques, electrophoresis, solvent extraction, gel filtration, and release of the radioactivity in a volatile form. Although some of the techniques used are time consuming, several procedures that are rapid and convenient have now been developed. For example, extraction of the end product into a solvent containing scintillator so that the extract can be counted directly by scintillation counting has been performed for at least two enzymes, monoamine oxidase (42) and diamine oxidase (43). DEAE-cellulose paper disks have been used extensively in the assay of certain kinases (44–46).

Samples of the kinases reaction mixture are filtered through the disks and the phosphorylated products are retained. Residual substrate is removed by washing and the disk is then counted by liquid scintillation counting.

In addition to the problems of separating substrate from product, there are other hazards in radiochemical techniques for enzyme assays. Where small concentrations of labeled substrates are used the specific activity may be considerably altered by the presence of endogenous substrates. False rates of reaction are therefore observed. Another problem is in the accurate determination of the specific activity of the unlabeled substrate, because the specific activity is the basis for calculating the amount of product formed.

The release of tritium from specifically labeled substrates has been used as the basis for the assay of several enzymes. These include amino acid hydroxylases (39, 47), synthetases (48, 49), and ribonucleotide reductase. Tritiated water can be recovered for counting by distillation if the substrate is not volatile (48) or by ion-exchange chromatography to remove labeled compounds (47). There are, however, several potential sources of error in this type of assay. The position, the stereospecificity, and the stability of the label must be accurately known if accurate rates of reaction are to be obtained. In general, high specific activity substrates tend to give high blanks because of the "leakage" of tritium into aqueous solution on storage. The tritium may also be released during the reaction by some component of the reaction mixture other than the enzyme being assayed so that the observed rate of reaction is greater than the true rate. Another source of error discovered at the National Institutes of Health is with the so-called "NIH-shift," in which tritium is retained in the product instead of being released as tritiated water. In the hydroxylation of aromatic acids there is an intramolecular migration of the tritium atom displaced by the hydroxyl group, which results in varying amounts of tritium being retained in the product instead of being released into solution (50). Apparent low rates of reaction are therefore observed but the quantitative significance of this effect seems to depend on the enzyme. The amount of tritium retained is small when tyrosine hydroxylase is the enzyme and by allowing the reaction to go to completion and determining the radioactivity retained in the product a correction factor can be applied. With phenylalanine hydroxylase, however, the amount of tritium is too high to allow the use of a simple correction (47).

In the estimation of unlabeled metabolites extracted from biological materials, enzymes have been used to convert nonradioactive substrates into labeled derivatives. A labeled cofactor is used in a manner similar to the use of a labeled reagent in isotope derivative analysis. A good example of such methods is the double-isotope enzymatic assay for histamine (51). Samples containing unknown amounts of histamine and tracer amounts of [^{3}H]histamine and [^{14}C]*S*-adenosylmethionine are incubated with a partially purified preparation of histamine methyltransferase from guinea pig brain. This enzyme is specific for the methylation of histamine (52). The product of the reaction, 1,4-methylhistamine, is extracted into chloroform and the ratio of $^{14}C/^{3}H$, determined by liquid scintillation counting, is directly proportional to the amount of unlabeled histamine present in the mixture. The use of a double-label technique does not require linearity of the enzymatic reaction because one label, the tritiated indicator, measures the amount of dilution by histamine in the sample, whereas the ^{14}C label of the *S*-adenosylmethionine monitors the degree of reaction. In this particular assay high specific activity tritiated histamine used as the indicator is not very stable and this may give rise to erroneous results (53), but the same principles have been used in the measurement of *S*-adenosylmethionine using another specific methylating enzyme, hydroxyindole-*O*-methyltransferase.

In summary, when a radioenzymatic assay is to be modified to achieve high sensitivity, Oldham (41) has made the following suggestions.

a. The labeled substrate should be used at the lowest specific activity compatible with its proposed use. Substrates supplied at high specific activity should be diluted with inactive substrate to reduce radiation decomposition.

b. The labeled substance should be stored under the best conditions to reduce chemical and radiochemical decomposition.

c. When tritiated substrates have been stored in aqueous solution, they should be lyophilized to remove any tritium that has exchanged with the solvent.

d. If necessary, the labeled substrate may be purified by using the same separation technique as that used in the enzyme assay. This can be especially helpful when blank values are too high.

e. If a concomitant nonenzymatic reaction is suspected because of a high blank, the effect of the buffer, pH, oxygen, trace metal ions, and complexing agent should be studied.

f. One should be aware that in certain instances the boiled enzyme extract used to determine the blank value can cause a nonenzymatic reaction that may mask the true enzymatic reaction. When this occurs, if appropriate, the use of the unnatural isomer as substrate may be advantageous.

g. The final volume for the desired reaction rate should be optimized, keeping in mind the sensitivity, reproducibility, and ease of sample preparation.

REFERENCES

1. *Chem. Eng. News* **49,** 25–26 (1971).
2. *Science* **167,** 237 (1970).
3. *Science* **168,** 1285 (1970).
4. Bayley, R. J., and Evans, E. A. "Storage and Stability of Compounds Labeled with Radioisotopes." Radiochemical Centre, Amersham, England, 1968.
5. Sommerville, I. F. "Determination of Steroid Hormones," Med. Monogr. No. 7. Radiochemical Centre, Amersham, England, 1970.
6. Yalow, R. S., and Berson, S. A. *Nature* (*London*) **184,** 1648 (1959).
7. Yalow, R. S., and Berson, S. A. *J. Clin. Invest.* **39,** 1157 (1960).
8. Ekins, R. P. *Clin. Chim. Acta* **5,** 453 (1960).
9. Ekins, R. P. *In* "*In vitro* Procedures with Radioisotopes in Clinical Medicine and Research," p. 325. IAEA, Vienna, 1969.
10. Murphy, B. E. P. *Nature* (*London*) **201,** 679 (1964).
11. Murphy, B. E. P. *In* "Steroid Assay by Protein Binding," (A. Diczkalusy, ed.), p. 37. Karolinska Institute, Stockholm, 1970.
12. Rothenburg, S. P. *Nature* (*London*) **206,** 1154 (1965).
13. Rothenberg, S. P. *Anal. Biochem.* **13,** 530 (1965).
14. Rothenberg, S. P. *In* "Radioisotopes in Medicine, *In Vitro* Studies" (R. L. Hayes, F. A. Goswitz, and B. E. P. Murphy, eds.), p. 413. U.S. At. Energy Comm., Oak Ridge, Tennessee, 1968.
15. Slaunwhite, W. R., and Sandberg, A. A. *In* "Steroid Assay by Protein Binding," p. 144. Karolinska Institute, Stockholm, 1970.
16. Berson, S. A., and Yalow, R. S. *Ciba Found. Colloq. Endocrinol.* [*Proc.*] **14,** 182 (1962).
17. Yalow, R. S., and Berson, S. A. *In* "Radioisotopes in Medicine, *In Vitro* Studies" (R. L. Hayes, F. A. Goswitz, and B. E. P. Murphy, eds.), p. 7. U.S. At. Energy Comm., Oak Ridge, Tennessee, 1968.
18. Yalow, R. S., and Berson, S. A. *In* "Principles of Competitive Protein-Binding Assays" (W. D. Odell and W. H. Daughaday, eds.), p. 1. Lippincott, Philadelphia, Pennsylvania, 1971.
19. Yalow, R. S., and Berson, S. A. *In* "Proteins and Polypeptide Hormones" (M. Margoulies, ed.), Excerpta Med. Found., Amsterdam, 1969. Int. Cover. Ser. No. 161, p. 71.
20. Ekins, R. P., Newman, G. B., and Riordan, J. L. H. *In* "Statistics in Endocrinology" (J. W. McArthur and T. Colton, eds.), p. 345. MIT Press, Cambridge, Massachusetts, 1970.
21. Ekins, R. P., and Newman, G. B. *In* "Steroid Assay by Protein Binding," p. 11. Karolinska Institute, Stockholm, 1970.
22. Ekins, R. P., Newman, G. B., and Riordan, J. L. H. *In* "Radioisotopes in Medicine, *In Vitro* Studies" (R. L. Hayes, F. A. Goswitz, and B. E. P. Murphy, eds.), p. 59. U.S. At. Energy Comm., Oak Ridge, Tennessee, 1968.

23. Ekins, R. P. *In* "Proteins and Polypeptide Hormones" (M. Margoulies, ed.), Int. Congr. Ser. No. 161, p. 672. Excerpta Med. Found., Amsterdam, 1969.
24. Ekins, R. P., and Newman, G. B. *Acta Endocrinol.* (*Copenhagen*) **64,** Suppl. 147, 11 (1970).
25. Midgley, A. R., Jr., Niswender, G. D., and Regan, R. W. *Acta Endocrinol.* (*Copenhagen*) **63,** Suppl. 142, 163 (1969).
26. Rodbard, D. and Weiss, G. H. *Anal. Biochem.* **52,** 10 (1973).
27. Ashcroft, J. *Anal. Biochem.* **37,** 268 (1970).
28. Litt, G. J. New England Nuclear Corp., Boston, Massachusetts (personal communication).
29. Sheppard, G., and Marlow, C. G. *Int. J. Appl. Radiat. Isotop.* **22,** 125 (1971).
30. Newsholme, E. A., and Taylor, K. *Biochim. Biophys. Acta* **158,** 11 (1968).
31. Brooker, G. A., Thomas, L. J., and Appleman, M. M. *Biochemistry* **7,** 4177 (1968).
32. Brooker, G. A., and Appleman, M. M. *Biochemistry* **7,** 4182 (1968).
33. Kobayashi, Y. *Anal. Biochem.* **5,** 284 (1962).
34. Reed, D. J. "Atomlight," No. 66, p. 1. New England Nuclear Corp., Boston, Massachusetts, 1968.
35. Oldham, K. G. "Radiochemical Methods of Enzyme Assay." Radiochemical Centre, Amersham, England, 1968.
36. Litwack, G., and Squires, J. M. *Anal. Biochem.* **24,** 438 (1968).
37. Leinweber, F. J. *Mol. Pharmacol.* **4,** 337 (1968).
38. Grahn, B., and Rosengren, E. *Brit. J. Pharmacol. Chemother.* **33,** 472 (1968).
39. Nagatsu, T., Levitt, M., and Undenfriend, S. *J. Biol. Chem.* **239,** 2910 (1964).
40. Oldham, K. G. *Biochem. Pharmacol.* **17,** 1107 (1968).
41. Oldham, K. G. *Int. J. Appl. Radiat. Isotop.* **21,** 421 (1970).
42. Otsuka, S., and Kobayashi, Y. *Biochem. Pharmacol.* **13,** 995 (1964).
43. Okuyama, T., and Kobayashi, Y. *Arch. Biochem. Biophys.* **95,** 242 (1961).
44. Sherman, J. R. *Anal. Biochem.* **5,** 548 (1963).
45. Newsholme, E. A., Robinson, J., and Taylor, K. A. *Biochim. Biophys. Acta* **132,** 338 (1967).
46. Furlong, N. B. *Anal. Biochem.* **5,** 515 (1963).
47. Guroff, G., and Abramowitz, A. *Anal. Biochem.* **19,** 548 (1967).
48. Lomax, M. I. S., and Greenberg, G. R. *J. Biol. Chem.* **242,** 109 (1967).
49. Kammen, H. O. *Anal. Biochem.* **17,** 553 (1966).
50. Guroff, G., and Daly, J. *Arch. Biochem. Biophys.* **122,** 212 (1967).
51. Snyder, S. H., Baldessarini, R. J., and Axelrod, J. *J. Pharmacol. Exp. Ther.* **153,** 544 (1966).
52. Brown, D. D., Tomchick, R., and Axelrod, J. *J. Biol. Chem.* **234,** 2948 (1959).
53. Kobayashi, Y., and Maudsley, D. V. *Anal. Biochem.* **46,** 85 (1972).

6

Practical Aspects of Sample Preparation

The primary objective of all sample preparative procedures is to obtain a homogeneous solution. Sample preparation remains very much an empirical procedure and adaptation of existing techniques to satisfy specific requirements is still within the domain of the individual investigator. This chapter provides an outline of some of the more established and promising procedures that have been developed in preparing samples of biological interest for counting. The problems associated with difficult samples can be overcome by the use of an automated combustion apparatus but such equipment is still rather rare and the investigator must usually rely on the traditional chemical methods for this purpose. In some of the original papers cited, the concentration of scintillators used seem, in the light of present knowledge, to be unnecessarily low and we have taken the liberty, therefore, of suggesting different formulations for the counting solution. We prefer bis-MSB to either POPOP or DMPOPOP as a secondary scintillator because of the greater flexibility offered by its higher solubility and, in some cases, butyl-PBD can be substituted for PPO to advantage. A section on processing acrylamide gels has been included because of the wide interest in processing gel slices for counting.

TISSUES

The most commonly used methods for preparing tissues involve either direct solution or combustion procedures. Direct solution is both simple and economical and should be tried first. Combustion techniques require special apparatus and further discussion is deferred to near the end of this chapter.

Perchloric Acid Method

This method can be used with standard laboratory reagents and is applicable to tissues, blood, urine, body fluids, and filter disks (1).

Up to 0.2 g of tissue is dispensed into a counting vial. An equal volume of 60% perchloric acid is added followed by 30% H_2O_2 in an amount equivalent to twice the initial volume of the sample. The solution is mixed and the vial capped to prevent evaporation losses. The sample is then heated in an oven at 70°–80°C for 30–60 minutes with occasional shaking. After cooling 5–6 ml of Cellusolve are added to the vial followed by 10 ml of a toluene–PPO cocktail. The PPO concentration is 6 g/liter.

The perchloric acid should be added first; otherwise, frothing prevents adequate mixing of the reagents. After heating, the vial contents should be clear and colorless. Sometimes a fine precipitate is observed but it is soluble in the final solution and has no effect on the counting efficiency. The use of a secondary scintillator, such as DMPOPOP or bis-MSB, should be avoided as they acquire a yellow color in the final solution and this causes substantial color quenching. Although the original data show little to no loss of radioactivity via $^{14}CO_2$ or tritiated water, care should be exercised to avoid excessive heating during the oxidation.

NCS Method

The use of this agent for solubilizing tissues has been extensively studied and the optimal sample size determined for some biological material for digestion by NCS is shown in Table 22 (2). The weight to volume ratio of sample to solubilizer that gives optimum conditions (as indicated by the highest figure of merit) varies considerably from tissue to tissue and even for the same tissue from different animals. Optimal sample composition is also influenced by the isotope used. Tritium samples are more severely quenched than carbon-14 samples. To offset this the volume of NCS used is reduced. Consequently, for tissues labeled with tritium the

TABLE 22

Optimal Sample Composition for Counting ^3H and ^{14}C in Wet Tissue[a]

Tissue	Form	Solvent	Solvent volume (ml)	Sample wet weight or volume	Percent efficiency	Figure of merit
^3H-Muscle	Coarsely ground	NCS	5.0	950 mg	18	17,000
		HyOH[b]	2.1	450 mg	14	6,300
	1:4 homogenate	NCS	7.6	280 mg	20	5,700
		HyOH	3.7	110 mg	10.6	1,170
^{14}C-Muscle	Coarsely ground	NCS	11.5	2300 mg	47	109,000
		HyOH	5.4	1150 mg	42	48,000
	1:4 homogenate	NCS	15.3	560 mg	58	32,200
		HyOH	7.0	210 mg	47	9,800
^3H-Liver	Coarsely ground	NCS	0.7	75 mg	18	1,350
		HyOH	0.4	40 mg	19.2	770
	1:4 homogenate	NCS	4.2	125 mg	19.2	2,400
		HyOH	2.0	60 mg	12.2	730
^{14}C-Liver	Coarsely ground	NCS	1.9	220 mg	42	9,200
		HyOH	1.3	130 mg	41	5,300
	1:4 homogenate	NCS	9.3	280 mg	51	14,300
		HyOH	4.5	135 mg	42	5,700
^3H-Blood	Whole	NCS	1.8	0.18 ml	18	3.23
		HyOH	0.7	0.11 ml	14.1	1.55
^{14}C-Blood	Whole	NCS	4.5	0.45 ml	45	20.2
		HyOH	1.5	0.25 ml	48	12.1
^3H-Plasma		NCS	8.0	1.2 ml	22	26.4
		HyOH	2.7	0.5 ml	16.4	8.2

[a] From Hansen and Bush (2). [b] Hyamine hydroxide.

optimal sample size is lower than with carbon-14 labeled tissues. The procedure for liver is described here.

Minced tissue (220 mg), labeled with carbon-14, is placed in a glass scintillation vial. Then 1.9 ml NCS are added and the tissue is digested at 35°–50°C until no trace of solid remains. For tissues labeled with tritium the sample weight is reduced to 75 mg and the volume of NCS to 0.7 ml. The temperature during digestion should not exceed 60°C as higher temperatures produce excessive quenching. When digestion is complete the sample is cooled to room temperature and diluted to 20 ml with a toluene–PPO–bis-MSB cocktail. The concentration of PPO is 6 g/liter and of bis-MSB 100 mg/liter. The sample should be cooled to the temperature of the counter for at least 30 minutes before it is counted.

Hyamine Method

As far as scintillation counting has been able to establish tradition, the use of Hyamine is traditional, although for many purposes better agents are now available. Nevertheless, it is a well proved and versatile reagent and still serves as the standard with which others are compared. The comments in the previous section on NCS are all applicable to Hyamine. As a representative example of the procedure, the method for preparing muscle labeled with tritium is given here.

Muscle, finely minced and weighing 400–450 mg, is placed in a counting vial and 2 ml of Hyamine hydroxide 10-X is added. The vial is capped and the contents are mixed. The sample is incubated with occasional agitation at 50°C until no trace of solid remains. This time may vary from one to several hours. When digestion is complete the vial is cooled and 10 ml of a counting solution containing 0.6% PPO and 0.01% bis-MSB in toluene is added. The vial is shaken and placed in the counter for at least 30 minutes before it is counted.

For carbon-14 labeled tissues the amounts of sample and Hyamine can be increased. For dry samples a small amount of water can be added to facilitate digestion.

Bio Solv Method

About 50 mg wet weight of tissue is weighed into a liquid scintillation vial, 1 ml of 2 *N* sodium hydroxide is added, and the sample is heated in an oven at 80°C for 1 hour. The volume should be maintained constant by adding a few drops of water. After digestion is complete, the sample mixture is cooled and enough solubilizer, BBS-2, is added to bring the pH to about 7. This is usually about 2–3 ml of BBS-2. The solution is mixed until clear. After 10 minutes two drops of a 4% solution of stannous chloride in 0.1 *N* hydrochloric acid is added to aid solubility. Ten milliliters of a toluene counting solution containing 0.8% butyl-PBD or 0.7% PPO and 0.05% bis-MSB is added to the sample. If a clear solution does not result more solubilizer should be added. Twenty minutes should be allowed for the samples to come to equilibrium.

The solubilizer can be mixed with the counting solution before the sample is added. When large numbers of samples are being routinely assayed, this results in a significant saving of time.

Highly colored samples should be decolorized by adding a few drops of 30% hydrogen peroxide or benzoyl peroxide. The water volumes should be kept to a minimum so that minimum volumes of the solubilizer are required.

Other Solubilizers

Nitric acid has been used for digesting tissues (3–5). Although digestion is more rapid than with other methods, the procedure should be strictly followed to avoid discoloration by nitric acid fumes and loss of radioactivity via $^{14}CO_2$ formation or the evaporation of tritiated water. Some early workers (6) have claimed equal success with KOH digestion as with Hyamine for a variety of tissues and biological specimens, but later work (2) has questioned the efficacy of KOH digestion and demonstrated the superiority of the quarternary amines, NCS and Hyamine, over KOH for solubilizing tissues.

Aquasol Method

Emulsion counting has become very popular in recent years. Although this technique has normally been used to accommodate hydrophilic compounds, it has been applied to proteins as well. Although emulsions and suspensions are difficult to standardize, properly controlled the technique can satisfy the criteria for useful counting. In the method outlined below, the authors (7) showed the linearity of the counts observed as a fraction of the volume of liver homogenate corresponding to a maximum of 83 mg of liver from animals injected with 1 or 5 μCi of [^{3}H]leucine.

Liver homogenate of 0.2-ml aliquots (1:2, w/v) in distilled water is added to 7 ml of a 7% TCA solution containing 4 mg/ml celite and 0.1 *M* leucine. The sample is heated at 95°C for 30 minutes and then filtered through a Millipore filter (SCWP 0 2500), washed twice with 10 ml of 5% TCA, and washed twice with 5 ml of 70% ethanol. The precipitate and filter paper are transferred to a 3-dram scintillation vial (minivial) and 2 ml of water are added, followed by 6.75 ml of Aquasol. The vial is capped, shaken vigorously, and counted. Under normal operating conditions, 20 animals can be killed and all samples prepared for counting in 2 hours.

Generally speaking, the quarternary amines NCS, Protosol, Soluene-100, Eastman tissue solubilizer, and Hyamine hydroxide can probably be substituted for one another in any procedure calling for their use. Whenever these solubilizers are used, with the exception of Protosol, butyl-PBD should not be used in the counting solution because a yellow color develops

(8, 9). Therefore, PPO is the scintillator of choice when quarternary amine tissue solubilizers are used.

BLOOD

Any of the above procedures outlined for tissues can also be used for digesting whole blood. Color, however, produces severe quenching and the samples should be decolorized by adding a few drops of benzoyl peroxide or hydrogen peroxide. The decolorizing agent is usually added during the alkaline digest. In the perchloric acid method this is not necessary because the inclusion of hydrogen peroxide is part of the standard procedure.

A procedure has been described for extracting substances from whole blood (10). The method is unusual in that the extraction is carried out with the scintillation solvent.

One to 2 ml of blood are placed in a plastic vial and shaken with 15 ml of the scintillation fluid. This solution contains 12 g PPO, 1 g bis-MSB, 750 ml dioxane, 125 ml anisole, 125 ml dimethoxyethane, and 1 mg/ml butylhydroxytoluene (BHT). The mixture containing the sample is shaken briefly at 15°–20°C. After about 30 minutes it is filtered through a cotton plug. The precipitate is then washed with the scintillation fluid (2×2 ml) and the combined filtrate is counted after 30 minutes. The addition of BHT serves two purposes. It prevents peroxide formation and reduces chemiluminescence. If the radioactivity in the coagulated cells is required the precipitate can be dried and combusted in an oxygen flask.

URINE

Small amounts of urine (up to 1 ml) can sometimes be directly incorporated into a dioxane solution containing 0.4% PPO, 0.01% bis-MSB, and 12% naphthalene (11). However, there is a tendency for urine specimens with high salt contents to precipitate out when added to dioxane solutions. An alternative procedure is to add one of the solubilizing agents followed by a toluene cocktail. It may be necessary to decolorize the samples. If the urine specimens contain precipitated protein, it is advisable to digest the sample as outlined for tissues.

Urine, 1 ml, can be conveniently counted in an emulsion system consisting of a 1:1 (v/v) mixture of toluene and Triton X-100 containing a final concentration of 0.5% PPO and 0.01% POPOP (12). Similarly, 3.5 ml of urine can be taken up in 11.5 ml of Aquasol, shaken vigorously, and counted as a gel.

PLASMA

Up to 1 ml of plasma can be incorporated into a toluene cocktail with the aid of a solubilizing agent, such as Hyamine, Protosol, NCS, Soluene-100, or BBS-3. Heating the sample with the solubilizing agent before adding the counting solution is not usually necessary.

Alternatively, 1 ml of plasma can be counted as an emulsion in 10 ml of a 1:1 mixture of toluene and Triton X-100 containing a final concentration of 0.5% PPO and 0.01% POPOP (12) or in any of the commercial emulsifiers, such as Aquasol or Instagel.

WATER

Small volumes of water, up to 1% of the final volume, can be incorporated into a toluene cocktail containing a suitable additive, such as ethanol or ethoxyethanol. Larger volumes can be added directly to a dioxane–naphthalene counting solution or mixed with a solubilizing agent and added to a toluene cocktail. Details of the composition of the counting solutions are contained in Table 4. An interesting and sometimes useful alternative is the use of Triton X-100 (13). A toluene counting solution and Triton X-100 are mixed in the ratio of 2:1 (v/v) and used directly for aqueous solutions. Up to 15% of aqueous solutions containing carbon-14 can be counted in this manner at similar efficiencies to those obtained in homogeneous solution (14). For tritium samples, particularly, close attention to experimental details is necessary if reliable results are to be obtained (15).

The performance data of commercial emulsifiers can be obtained from the manufacturers. An inexpensive, high-capacity emulsifier has recently been reported (16), consisting of 25% Triton X-114 in xylene that contains 0.3%, (w/v) PPO.

PROTEINS

Proteins can be conveniently treated by any of the methods used for tissues. They can also be combusted as outlined in the section "Nonautomatic Combustion Techniques." However, the preparation of proteins for scintillation counting is often combined with an analytical procedure. Numerous variations are possible. For example, proteins can be separated by paper chromatography and strips containing the designated protein placed in a scintillation vial and a toluene cocktail added. Alternatively, the protein can be precipitated onto a suitable support, such as glass fiber disks, and then counted. As with all methods involving a support medium, nonreproducible results may occur because of partial elution from the support by the scintillator solvent. Variations in counting efficiencies may also result from an uneven distribution of the material over the support medium giving rise to differences in self-absorption. This can sometimes be a major problem for tritium samples. Preparation of proteins separated on acrylamide gels for counting is discussed separately below.

AMINO ACIDS

Small amounts of amino acids in aqueous solution can often be added directly to a toluene–ethanol mixture, such as solution B in Table 4. If the amino acid precipitates, digestion with an organic solubilizing agent may be necessary. Up to 25 mg of most amino acids can be counted in 10 ml of a counting mixture of toluene and Triton X-100 (2:1, v/v). The system can be used at low temperatures and the samples are stable over an extended period of time (14).

LIPIDS

Many lipids can be incorporated directly into a toluene counting solution. If this is unsuccessful then the compound can be combusted or digested as outlined for tissues.

Scrapings from a thin-layer plate can be suspended in 15 ml of a toluene–PPO–bis-MSB solution containing 4% Cab-O-Sil. If phosphotides are being assayed, polyethylene vials should be used, for adsorption on the surface of glass vials can occur. For fatty acids, the silica scrapings should be acid; when acetic acid is not used in the TLC separation, 50 μl of acid should be added directly to the counting vial. If adsorption of long-chain fatty acids on silica gel is allowed to occur in the vial, a decrease in counting efficiency caused by self-absorption may occur (17).

Gels prepared from Cab-O-Sil, although relatively efficient, are awkward to handle. An alternative method employs toluene diisocyanate (TDI) and a branched aliphatic primary amine, such as Armeen L-11 (18).

To 10 ml of a toluene–PPO–bis-MSB counting solution is added 0.17–0.2 ml of Armeen L-11. The sample, weighing up to 1 g, is then added, followed by 0.05 ml TDI. The reaction requirement is 2 moles of the amine to 1 mole of TDI and because excess TDI lowers the viscosity of the gel and increases quenching it is advisable to add a slight excess of the amine. Sharp swirling of the solution starts the gelation and simultaneously suspends the particulate matter. The gel sets in a few seconds to a few minutes. The gel is clear and preparation of a large number of samples is easy. Counting efficiencies are reported to be as good as for Cab-O-Sil.

It is reported that up to 300 mg of silica gel scrapings from a thin-layer plate used for lipid analysis can be counted directly in 15 ml of Aquasol without impairment of counting efficiency (19). Another interesting method is to use hydrofluoric acid to dissolve the silica gel, leaving the lipid unaffected, and then to solubilize the digest in a toluene–Triton X-100 counting solution (20).

CARBOHYDRATES

Soluble carbohydrates may be incorporated into a toluene cocktail containing a blending agent, such as ethanol or ethoxyethanol, or by using a dioxane–naphthalene solution. Insoluble material may be combusted or digested and solubilized into a toluene cocktail for tissues.

STEROIDS

Small amounts of steroids are usually incorporated directly into a toluene–PPO–bis-MSB solution. More polar compounds are dissolved in 0.2 ml of methanol before the counting solution is added. Steroid conjugates are dissolved in a minimum amount of water, followed by 3 ml of ethanol and 10 ml of the toluene cocktail. With more than a few milligrams of material, recourse to a dioxane cocktail may be necessary.

BACTERIA

Any of the methods described for tissues may be used. Formamide is reported to be superior to Hyamine for digesting bacterial cells (21). One milliliter of formamide per 5–20 mg of bacterial cells is added to a glass-stoppered test tube and digested at 60°C for 2 hours. An aliquot of the digest containing approximately 10 mg dry weight of cells is then transferred to a counting vial and 10 ml of a toluene–PPO (3 g/liter) counting solution and 5 ml of ethanol are added. For smaller amounts of cells (6–7 mg) autolysis in water for 2 hours at 70°C should be sufficient to bring the material into solution. The amount of water used is approximately 1 ml for 30 mg of cells. Autolysis is not recommended for larger amounts of cells. Color quenching complicates the use of Hyamine and dioxane–naphthalene solutions give lower counting efficiencies than toluene cocktails.

NUCLEIC ACIDS

Nucleic acids can be treated as described for tissues. Alternatively, they can be prepared by heating with concentrated hydrochloric acid (22). Up to 10 mg of RNA or 3 mg of DNA is placed in a counting vial and 0.5 ml of water or 0.2 ml of 2 *N* NaCl is added. After concentrated HCl (0.04 ml) is added, the sample is incubated at 50°C for 20–24 hours in silicone rubber-stoppered vials. Hyamine (0.1 ml) or its equivalent is added after incubation, followed by 10 ml of a counting solution consisting of 100 g naphthalene, 10 g PPO, 500 mg bis-MSB, and dioxane to make 1 liter.

A more recent method that is reported to give high counting efficiencies for tritium utilizes the solubility of the cetyltrimethylammonium salt of nucleic acids in 2-methoxyethanol and ethanol (23).

Ten milliliters of a solution containing 0.1 mg/ml carrier RNA and 0.05 *M* NaCl/0.005 *M* phosphate buffer, pH 6.7, is added to each 3 ml of sample. Then 1 ml of cetyltrimethylammonium bromide is added and the fractions are left at 0°–4°C overnight to ensure complete precipitation. The precipitate is then collected onto a glass fiber filter and washed with distilled water to remove residual salt and detergent. The filters are then dried overnight at room temperature or at 40°–50°C for 2 hours. The filters are then placed in a scintillation vial, 1 ml of methoxyethanol is added, and the solution is boiled for 30 seconds on a hot plate. Then 15 ml of toluene cocktail containing 0.4% PPO and 0.01% bis-MSB are added and the glass fiber is dispersed by vigorous shaking for 5 seconds. The sample is then counted.

The precautions concerning the use of glass fiber filters in counting samples as well as other potential pitfalls are discussed in a review of labeled nucleotides in biochemistry (24).

$^{14}CO_2$ FROM A WARBURG FLASK

This procedure is used for trapping $^{14}CO_2$ produced in a reaction carried out in a Warburg flask. The method is generally applicable to any procedure in which $^{14}CO_2$ is produced in a closed system. The method as currently used in this laboratory is described here.

A strip of Whatman No. 1 filter paper containing 0.04 ml of Hyamine or its equivalent is placed in the center well of a Warburg flask. The filter strip is edged with paraffin wax to prevent the Hyamine from running onto the walls of the center well. The reaction is carried out in the usual way and is stopped by adding 0.2 ml of 1 *M* citric acid to the side arm. If a rubber multidose stopper is substituted for the more conventional glass stopper to seal the side arm, this can easily be done with the aid of a hypodermic needle. After tipping the acid into the main vessel the flask is shaken for a further 30–60 minutes to remove dissolved $^{14}CO_2$. The filter strip is then removed and added to a counting cocktail containing 7 parts toluene to 3 parts ethanol (v/v) 0.6% PPO, and 0.1% bis-MSB.

Various modifications of this procedure have been described (25, 26). Levine and Watts (27) describe a method that uses a polyethylene counting vial as the incubation vessel. A straightened wire paperclip with a narrow loop on one end is forced through the center of the cap of a polyethylene vial so that the loop end projects from the lower part of the cap. A rectangular piece of Whatman 3 mm filter paper measuring 1×3 cm is rolled into a cylinder and clamped firmly into the loop. The paper is dipped in Hyamine and allowed to drip dry. The loop should be so positioned that when the cap is screwed onto the vial the filter paper is entirely in the upper third of the vial and does not touch the edges of the vial. The incubation is carried out with the vial cap screwed on tightly. To terminate the reaction 2 ml of 6 *N* HCl is injected through the side of the vial with a 23 gage disposable needle; the hole in the vial is sealed promptly with adhesive tape. Incubation is continued for 30 minutes after acidification to permit quantitative adsorption of CO_2 onto the filter paper. The vial is then opened and the cap and filter paper are transferred to another counting vial containing 10 ml of solution D (see Table 4). The vial cap assembly may be reused after acid washing.

A procedure for counting barium [^{14}C]carbonate has been described in detail by Cluley (28). Dried $Ba^{14}CO_3$ is ground in a glass mortar and an aliquot between 0.1 and 0.5 g is weighed. The fine powder is then added to a counting vial containing 400 ± 3 mg Cab-O-Sil. Ten milliliters of a toluene–PPO counting solution are then added and the mixture is shaken vigorously for a few minutes. For determining counting efficiency a barium carbonate standard should be used.

NONAUTOMATIC COMBUSTION TECHNIQUES

The most established method is that described by Davidson and Oliverio (29). A 2-liter Erlenmeyer filter flask is used as the combustion vessel. The sample holder consists of a basket made from platinum or nichrome attached to a wire stalk, which is carried by a Pyrex glass rod through a rubber stopper in the neck of the flask. Neoprene stoppers are used because they are more resistant to scintillation solvents than are ordinary rubber or vinyl stoppers. Closure of the side arm of the flask is effected with a 2-inch length of 5.0-mm o.d. $\times$ 1.6-mm wall silicone rubber tubing and a pair of standard forceps. The jaws of the forceps are covered with rubber sleeves to prevent cutting of the silicone tubing. An infrared light source is used for igniting the sample.

The sample, weighing up to 250 mg dry weight, is placed in a small cellophane bag, which may be obtained commercially. The sample is dried and placed in the wire mesh holder and the flask is flushed with oxygen. If the sample size is small or high in inorganic salts, fuel in the form of a sucrose solution or a small piece of filter paper should be added to the sample to achieve satisfactory combustion. The sample is marked with black ink to further facilitate combustion. The infrared light source is then directed on the sample and combustion is normally complete within 1 minute. The flask is cooled for 5 minutes and the counting solution is added. For tritium the counting solution consists of 30% methanol, 0.6% PPO, 0.1% bis-MSB, and toluene. The clamp on the side arm of the flask is released, 15 ml of this solution is pipetted through the side arm, and the clamp is replaced. The flask is swirled gently and placed in a cooling bath at $-15°C$ or lower with the cooling confined to the bottom inch of the flask. After 15–20 minutes an additional 3 ml of counting solution is added to remove any solution trapped in the side arm of the flask. The contents of the flask are again mixed and a 15 ml aliquot is removed for counting. For collecting carbon dioxide the counting solution should consist of 27% ethanol, 27% phenylethylamine, 0.6% PPO, 0.1% bis-MSB, and toluene. The phenylethylamine should be redistilled before use. The procedure for adding this solution is the same as that just discussed.

For small amounts of material (less than 3 mg) a method that uses a counting vial as the combustion flask has been described by Gupta (30).

The combustion apparatus consists of a low-potassium glass counting vial, a platinum–iridium coil stand, and a paper cup. About 10–12 cm of platinum–iridium wire (gauge 24) is used to make one coil stand. This is done by turning the wire three to four times around a glass rod (4 mm o.d.) at its tapered end and then finishing with one single loop 1.5 cm in diameter. The paper cups, weighing 2 mg, are prepared from cut pieces of either silicone-treated eyeglass tissue or optical lens paper by pushing the center with the blunt end of a rubber stopper. The tip of the cup is blackened with a marking pencil to facilitate ignition. In each cup is placed a small cotton pellet to serve as a fuel and absorbent for liquid material. Glass fiber disks, containing 0.1 ml phenylethylamine, are used directly inside the vial for direct absorption of the combustion products of carbon-14 and sulfur-35. Ignition is triggered with a 150-W projection lamp.

Dry tissue, up to 3 mg, is placed in the cup mounted on the platinum coil. Wet tissue (10–15 mg) or liquid materials (up to 100 μl) are placed in the cups and dried before ignition. The lamp is focused on the black part of the cup and the sample is ignited. Combustion is complete within a few seconds. The vial is then cooled briefly in liquid nitrogen and 15 ml of a suitable counting solution such as toluene–ethanol–PPO–bis-MSB is then added. The method is reported to give good efficiencies for tritium, carbon-14, and sulfur-35.

POLYACRYLAMIDE GELS

Polyacrylamide gel electrophoresis is a technique widely used for the separation and identification of biological materials. The location of the various compounds within the gel is determined by staining and by the presence of radioactive markers in discrete zones. Analysis of the radioactive content of polyacrylamide gels has presented a problem because they are insoluble in the common solvents used in liquid scintillation counting.

One method for solubilizing gel slices is to treat 1- to 2-mm slices of the gel with about 0.5 ml of 30% hydrogen peroxide and to digest the gel for varying times at about 50°C. The hydrogen peroxide also decolorizes the stained gels. Some authors use a 1-hour digestion at 55°C (31), overnight at 40°C (32), and overnight at 60°C (33). Once the gel is dissolved and decolorized, the digest is taken up with between 5 and 15 ml of Aquasol for counting. High counting efficiency for both tritium and carbon-14 is reported. The digestion time is a function of the material being treated. Most reports in which hydrogen peroxide has been used have dealt with various RNA fractions, virus, and enzymes. Goodman and Matzura (34) have criticized the digestion of gels with hydrogen peroxide and heat on the grounds that radioactive carbon dioxide and water may be lost. They proposed the digestion of 1-mm gel slices with 0.25 ml of a mixture containing 1 part concentrated ammonium hydroxide and 99 parts 30% hydrogen peroxide at room temperature for 4–8 hours. The clear digest is then taken up in 10 ml of counting solution consisting of 1 part Beckman BBS-3 and 5 parts toluene-based counting solution. This procedure has been used for gels containing amino acids, peptides, and proteins and is free from the danger of loss of radioactivity because of low-temperature solubilization.

Solubilization of RNA for liquid scintillation counting has been reviewed by Adesnik (35). RNA can be hydrolyzed to nucleotides with dilute alkali, which can easily diffuse out of the gel and be dissolved in an appropriate aqueous counting solution. For example, slices of a BIS cross-linked gel can be treated with 0.5 ml of 2 *N* NH_4OH for several hours at 50°C and then the sample is dried. Water (0.5 ml) is added and the sample taken up in a dioxane-based counting solution. Alternately, either the BIS or the ethylene diacylate cross-linked gels can be dissolved in 0.7 ml of a mixture of 1 part water and 9 parts Protosol (or equivalent) in a tightly capped scintillation vial, which is heated overnight at 50°C. The digest is taken up in a toluene-based counting solution (36–38). Caution against chemiluminescence must always be observed when dealing with basic digests, as discussed in Chapter 3.

Although there is sparse data at this writing, gel samples can be ideally prepared for solution counting of tritium and/or carbon-14 by combustion using one of the automated commercial apparatus. Good results have been reported for the Schoniger technique (39) applied directly in a counting vial, but combusting individual gel samples using this technique is tedious and not amenable to a large number of samples. Compared to digestion procedures, automated combustion offers a rapid and accurate means of preparing samples free from the chemiluminescence problems associated with most acrylamide gel counting procedures.

A novel way to count acrylamide gels directly has been devised by Maizel (40, 41). A special apparatus expresses a gel through a fine orifice, which is met by a stream of counting solution, and both are directed into a series of counting vials on a fraction collector. ^{32}P-labeled gel samples have been prepared by this method and counted by Cerenkov counting (42). Maizel's procedure has been adapted for continuous flow counting by Bakay (43), who reports a counting efficiency for tritium of 30% with this modification.

The complete solution of the gel is apparently not required for satisfactory counting. This is indicated by the observation that when acrylamide gels are treated with a solubilizer, such as NCS or its equivalent, the gel swells rapidly and the entrapped macromolecules (proteins), which are presumably at least partially hydrolyzed by the strong organic base,

Sodium hydroxide is sometimes used for collecting carbon dioxide and the resultant solution is suspended in Cab-O-Sil and counted in a toluene cocktail. This is a useful procedure for *in vivo* collection of respiratory carbon dioxide.

Although the use of aromatic solvents has presented difficulties in preparing some types of samples, it provides an opportunity for combining the scintillation technique with an analytical procedure. An example of this is in certain enzyme assays when the substrate and product can be separated by extraction into a solvent, such as toluene, used as a base for a counting cocktail. If the solvent contains scintillator the extract can be counted directly. Such assays have been described for steroid sulfatase (48), diamine oxidase (49), monoamine oxidase (50), and proteolytic enzymes (51). Separation can be further facilitated by freezing the aqueous phase and pouring the upper organic layer into a scintillation vial. These methods are simple and rapid and reduce the losses incurred by transferring with pipettes.

REFERENCES

1. Mahin, D. T., and Lofberg, R. T. *Anal. Biochem.* **16**, 500 (1966).
2. Hansen, D. L., and Bush, E. T. *Anal. Biochem.* **18**, 320 (1967).
3. O'Brien, R. D. *Anal. Biochem.* **7**, 251 (1964).
4. Eldefrawi, M. E. *Anal. Biochem.* **17**, 353 (1967).
5. Pfeffer, M., Weinstein, S., Gaylord, J., and Indindoli, L. *Anal. Biochem.* **39**, 46 (1971).
6. Oliverio, V. T., and Davidson, J. D. *J. Pharmacol. Exp. Ther.* **137**, 76 (1962).
7. Friedman, M. A., Millar, G., McEvoy, A., and Epstein, S. S. *Anal. Chem.* **43**, 780 (1971).
8. Dunn, A. *Int. J. Appl. Radiat. Isotop.* **22**, 212 (1971).
9. McEvoy, A. F., Dyson, S. R., and Harris, W. G. *Int. J. Appl. Radiat. Isotop.* **23**, 338 (1972).
10. Houtman, C. A. *Int. J. Appl. Radiat. Isotop.* **16**, 65 (1965).
11. Butler, F. E. *Anal. Chem.* **33**, 409 (1961).
12. Chapman, D. I., and Marcroft, J. *Int. J. Appl. Radiat. Isotop.* **22**, 371 (1971).
13. Patterson, M. S., and Greene, R. C. *Anal. Chem.* **37**, 854 (1965).
14. Turner, J. C. "Sample Preparation for Liquid Scintillation Counting," Review 6. Radiochemical Centre, Amersham, England, 1967.
15. van der Laarse, J. D., *Int. J. Appl. Radiat. Isotop.* **18**, 485 (1967).
16. Anderson, L. E., and McClure, W. O. *Anal. Biochem.* **51**, 173 (1973).
17. Snyder, F., and Stephens, N., *Anal. Biochem.* **4**, 128 (1962).
18. Bollinger, J. N., Mallow, W. A., Register, J. W., and Johnson, D. E. *Anal. Chem.* **39**, 1508 (1967).
19. Kritchevsky, D., and Malhotra, S. *J. Chromatogr.* **52**, 498 (1970).
20. Shaw, W. A., and Harlan, W. R. *Anal. Biochem.* **43**, 119 (1971).

diffuse out into the liquid phase in which they are readily counted (37, 4
After 10–20 hours of digestion with solubilizer at 33°–45°C of 1-mm sli
of gel, the undissolved gel is removed after being counted and no p
ceptible loss of either tritium or carbon-14 counts or changes in the isotc
ratio of the samples have been observed (37). Birnboim (45), who descrit
a semiautomatic gel slicing apparatus, has also found that complete soluti
of the gel by an organic solubilizer is not required for good counting. Pa
(46), however, has evaluated a number of commercial organic amines
gel solubilizing agents and cautions against partial dissolution of the g
recommending complete solution of the gels for best counting results.

An emulsion counting system with high solubilizing power, especial
for acrylamide gel in which DATDA is substituted for BIS as the cros
linking agent, has been reported (16). It is a Triton X-114–xylene mixtu
(1:3, v/v) containing 0.3% PPO (w/v) as scintillator. The gel is dissolve
in 0.5 ml of 2% periodic acid in 2 hours at room temperature and the clea
digest taken up in 5 ml of this cocktail. Good recoveries and high countin
efficiencies are reported for both tritium and carbon-14. However, th
scintillator concentrations recommended by the authors are not optima
and should be increased to about 0.6% (w/v).

MISCELLANEOUS

It is clear that any one compound can often be treated in a variety of different ways, as dictated by circumstances and inclination. Amino acids, for example, can be counted in suspension or spotted onto filter paper, dried, and placed in a scintillation vial. From the standpoint of cost and availability, the preferred solvent for such heterogeneous systems is toluene but the main impurity of toluene is water and this may cause partial elution of the support. This leads to nonreproducible results and presents difficulties in determining absolute counting efficiency.

An adaptation of counting in a heterogeneous system has been described that markedly reduces the amount of counting solution required (47). The sample is impregnated on a glass fiber medium and dried under an infrared lamp. The strip is then wetted with a few drops (0.2–0.5 ml) of scintillator fluid, sealed in a plastic bag, and counted in a plastic vial. The decrease in sample volume reduces the background count and the cost of preparing the sample.

Alcoholic KOH has been used for digesting tissues but there is some doubt about its effectiveness (2). As there is an abundance of alternative procedures its continued use is not encouraged.

21. Neujahr, H. Y., and Ewaldsson, B. *Anal. Biochem.* **8,** 487 (1964).
22. Hattori, T., Aoki, H., Matsuzaki, I., Maruo, B., and Takahashi, H. *Anal. Chem.* **37,** 159 (1965).
23. Trewavas, A. *Anal. Biochem.* **21,** 324 (1967).
24. Monks, R., Oldham, K. G., and Tovey, K. C. "Labeled Nucleotides in Biochemistry," Review 12. Radiochemical Centre, Amersham, England, 1972.
25. Leinweber, F. J., and Walker, L. A. *Anal. Biochem.* **21,** 131 (1968).
26. Baggiolini, M., and Bickel, M. H. *Anal. Biochem.* **14,** 290 (1966).
27. Levine, R. J., and Watts, D. E. *Biochem. Pharmacol.* **15,** 841 (1966).
28. Cluley, H. J. *Analyst* **87,** 170 (1962).
29. Davidson, J. D., and Oliverio, V. T. "Atomlight," No. 60. New England Nuclear Corp., Boston, Massachusetts, 1967.
30. Gupta, G. N. *Anal. Chem.* **38,** 1356 (1966).
31. Diener, E., and Paetkau, V. H. *Proc. Nat. Acad. Sci. U.S.* **69,** 2364 (1972).
32. Benjamin, W. B. *Nature (London)* **234,** 18 (1971).
33. Dion, A. S., and Moore, D. H. *Nature (London)* **240,** 17 (1972).
34. Goodman, D., and Matzura, H. *Anal. Biochem.* **42,** 481 (1971).
35. Adesnik, M. *In* "Methods in Virology" (K. Maramorosch and H. Koprowski, eds.), Vol. 5, p. 125. Academic Press, New York, 1971.
36. Zaitlin, M., and Hariharasubramanian, V. *Anal. Biochem.* **35,** 296 (1970).
37. Terman, S. *Proc. Nat. Acad. Sci. U.S.* **65,** 985 (1970).
38. Udem, S. A., and Warner, J. R. *J. Mol. Biol.* **65,** 227 (1972).
39. McEwen, B. S. *Anal. Biochem.* **25,** 172 (1968).
40. Maizel, J. V. *Science* **151,** 988 (1966).
41. Maizel, J. V. *In* "Methods in Virology" (K. Maramorosch and H. Koprowski, eds.), vol. 5, p. 179. Academic Press, New York, 1971.
42. Clausen, T. *Anal. Biochem.* **22,** 70 (1968).
43. Bakay, B. *Anal. Biochem.* **40,** 429 (1971).
44. Basch, R. S. *Anal. Biochem.* **26,** 184 (1968).
45. Birnboim, H. C. *Anal. Biochem.* **29,** 498 (1969).
46. Paus, P. N. *Anal. Biochem.* **42,** 372 (1971).
47. Gupta, G. N. *Anal. Chem.* **39,** 1911 (1967).
48. Burstein, S., and Dorfman, R. I. *J. Biol. Chem.* **238,** 1656 (1963).
49. Okuyama, T., and Kobayashi, Y. *Arch. Biochem. Biophys.* **95,** 242 (1961).
50. Otsuka, S., and Kobayashi, Y. *Biochem. Pharmacol.* **13,** 995 (1964).
51. Roffman, S., Sanoka, U., and Troll, W. *Anal. Biochem.* **36,** 11 (1970).

Appendix

LICENSING REQUIREMENTS

General Use

The use and possession of radioisotopes are regulated in most areas of the world and the user must be either certified or licensed by the appropriate regulatory agency. In addition, the user may be subject to local and institutional regulations. Therefore, the investigator using isotopes should familiarize himself with all regulations because compliance with the law is his responsibility.

Federal Regulations, United States. Within the United States, the use and possession of most radioisotopes are governed by regulations published in the Federal Register issued by the United States Atomic Energy Commission.

State Regulations, United States. The Atomic Energy Commission has transferred its regulatory authority to states, which have established agencies for the control of radioactive materials. States that have their own regulatory agencies are Alabama, Arizona, Arkansas, California, Colorado, Florida, Georgia, Idaho, Kansas, Kentucky, Louisiana, Maryland, Mississippi, Nebraska, Nevada, New Hampshire, New York, North Carolina, North Dakota, Oregon, South Carolina, Tennessee, Texas, and Washington. Information regarding the name and address of the appropriate regulatory agency in any of these states may be obtained by writing

United States Atomic Energy Commission
Division of Licensing and Regulations
Washington, D.C. 20545

Although the procurement of small amounts of radioisotopes may be exempted from specific licensing by state and federal authorities, most vendors of radiochemicals only sell to licensed individuals.

Human Use of Radioisotopes

Radioactive chemicals purchased from vendors are not pharmaceutically refined and, therefore, not intended for use in humans. The suitability and safety of any radioisotope administered to humans is the sole responsibility of the investigator. Radioisotopes can be administered to humans only by a physician licensed to do so by the United States Atomic Energy Commission or its state equivalent.

Researchers in the United States who employ radioisotopes in humans must hold an "AEC-313A license for Non-Routine Medical Uses of By-Product Material" or an AEC Board License (medical uses). If the radioisotope is not under the jurisdiction of the AEC, i.e., cyclotron produced isotopes, then the investigator must provide evidence that he has filed a "Notice of Claimed Investigational Exemption for a New Drug" with the United States Food and Drug Administration.

In certain instances, the labeled compound to be administered to humans may be classified as a narcotic or a dangerous drug by the United States Department of Justice. The product description of these materials are listed by the vendor as requiring a "BNDD" form. The investigator must be registered with the Bureau of Narcotics and Dangerous Drugs (BNDD) and copies 1 and 2 of the investigator's BNDD order form must be supplied to the vendor of the radioactive drug prior to its shipment. The quantity of the drug ordered must be specified in units of weight and not in units of radioactivity. Information regarding BNDD regulations can be obtained from

United States Department of Justice
Bureau of Narcotics and Dangerous Drugs
P.O. Box 28083, Central Station
Washington, D.C. 20005
Attention: Registration Section

SOME USEFUL FORMULAS

Radioactive Decay

$$A = A_0\, e^{-\lambda t} = A_0\, e^{-0.693t/T_{1/2}}$$

$$\lambda = 0.693/T_{1/2}$$

where A_0 = activity, in any convenient units, at time zero
A = activity remaining after a time interval equal to t
λ = decay constant for the radioactive element derived using the same units as $T_{1/2}$
$T_{1/2}$ = half-life in a convenient unit of time
t = elapsed time in same units as $T_{1/2}$
e = base of natural logarithms, 2.7183

Specific Activity

Specific activity = $\lambda N = 0.693N/T_{1/2}$ = disintegrations per second per gram

where N = number of atoms per gram
$T_{1/2}$ = half-life of radioactive element in seconds

then

$$\text{Specific activity} = \frac{\lambda N}{3.7 \times 10^{10}} = \frac{N \times 1.873 \times 10^{-11}}{T_{1/2}} = \text{curies per gram}$$

Range of Beta Particle

For $0.01 \leq E \leq 2.5$ MeV

$$R = 412^{(1.265-0.0954 \ln E)}$$

$$\ln E = 6.63 - 3.2376\sqrt{10.2146 - \ln R}$$

where R = range (in mg/cm^2)
E = maximum energy (in MeV)

Internal Radiation Dosage

Biological Half-Life

$$T_b = \frac{0.693}{\lambda_b}$$

where λ_b = biological rate of elimination constant
T_b = biological half-life

Effective Half-Life

$$T_{\text{eff}} = \frac{(T_{1/2})(T_b)}{T_{1/2} + T_b}$$

where T_{eff} = effective half-life
$T_{1/2}$ = radioactive half-life (physical property)
T_b = biological half-life

Absorbed Beta Dose over a Period of Time

$$D = 73.8 E_\beta T_{\text{eff}} \frac{C_0}{M} (1 - e^{-\lambda_{\text{eff}} t})$$

$$\lambda_{\text{eff}} = \lambda_p + \lambda_b$$

where: D = dose (in rads)
E_β = average beta energy (in MeV) (tritium = 0.006; carbon-14 = 0.054
C_0 = initial concentration of isotope (in μCi)
M = mass of organ or whole body (in g)
T_{eff} = effective half-life (in days)
λ_{eff} = effective decay constant
λ_p = physical decay constant (fractional loss per day from radioactive decay)
λ_b = biological decay constant (fractional loss per day from biological elimination)
t = elapsed time (in days)

The factor 73.8 converts the units (μCi) (MeV/g) (day) to rads as follows

$$(3.2 \times 10^9) \left(\frac{1}{62.5 \times 10^6} \right) (1.44) = 73.8$$

The 1.44 converts half-life to average life.

Units of Radioactivity

One curie = 2.22×10^{12} disintegrations per minute = dpm
= 3.7×10^{10} disintegrations per second = dps
= 10^3 millicuries = 10^3 mCi
= 10^6 microcuries = 10^6 μCi
= 10^9 nanocuries = 10^9 nCi
= 10^{12} picocuries = 10^{12} pCi

Calculate Weight per Microcurie (Based on Highest Activity Theoretically Possible)

$$W = C(N)T_{1/2}(MW)$$

where
W = weight in micrograms
N = number of microcuries
$T_{1/2}$ = half-life
MW = molecular weight of material
C = constant

Values for C differ for various time units of $T_{1/2}$, as shown in the following tabulation.

C	$T_{1/2}$
8.87×10^{-14}	seconds
5.32×10^{-12}	minutes
3.19×10^{-10}	hours
7.66×10^{-9}	days
2.80×10^{-6}	years

HOW TO MAKE UP A RADIOACTIVE SOLUTION TO THE DESIRED SPECIFIC ACTIVITY

The first requirement in doing any radioactive analysis is to make up the radioactive reagent to the desired specific activity. Many reagents are of higher specific activity than is usually necessary and should be diluted before use. The following discussion describes, in detail, the preparation of [^{14}C]tyramine solution used in a radioenzymatic assay for the enzyme monoamine oxidase as an example of how reagents can be prepared to a required specific activity.

The radioactive tyramine is always purchased on the basis of its radioactive content, that is, on the total number of microcuries or millicuries desired. A microcurie is defined as that amount of radioactivity which contains 2.2×10^6 decompositions per minute (dpm). Therefore, 50 μCi of tyramine contains 50 μCi of radioactivity but different batches do not always contain the same weight (mg) of material. Fifty microcuries contain a total of $50 \times 2.2 \times 10^6$ dpm or 1.1×10^8 dpm.

The total amount or weight of tyramine containing 50 μCi of activity depends on its specific radioactivity. The specific radioactivity is defined as units of radioactivity per unit weight of material. Usually the activity for ^{14}C compounds is expressed as millicuries per millimole of compound. The specific activity expressed in the following forms are all equivalent.

$$\text{Curies/mole} = \text{millicuries/millimole} = \text{microcuries/micromole}$$

If the specific activity of a sample of tyramine hydrobromide is given as 4.35 mCi/mmole and the radioactivity per microgram of tyramine is desired, the calculations can be made as follows.

$$\text{Molecular weight of tyramine HBr} = 218 \text{ units (gm, mg, } \mu\text{g)}$$

$$\text{Molecular weight of tyramine} = 137 \text{ units}$$

$$\text{Specific activity} = 4.35 \text{ mCi/mmole} = 4.35\ \mu\text{Ci}/\mu\text{mole}$$

$$\text{Specific activity} = \frac{(4.35\ \mu\text{Ci})\ (2.2 \times 10^{6}\ \text{dpm}/\mu\text{Ci})}{137\ \mu\text{g tyramine}} = \frac{9.57 \times 10^{6}\ \text{dpm}}{137\ \mu\text{g tyramine}}$$

$$\text{Specific activity} = \frac{6.98 \times 10^{4}\ \text{dpm}}{\mu\text{g tyramine}}$$

Therefore, a sample of tyramine hydrobromide with a specific activity of 4.35 mCi/mmole contains 69,800 dpm/μg of tyramine. It is important to note that the calculation was based on the weight of the free tyramine, not tyramine hydrobromide. When the specific activity is reported for a unit weight of material, it is usually given for a unit weight of the free base (in this case, tyramine) and not its salt (in this case, tyramine hydrobromide).

The solution required is 5 μg of [^{14}C]tyramine per 0.1 ml. This is equivalent to 50 μg tyramine per milliliter. Because we know that the sample of [^{14}C]tyramine contains a total of 50 μCi with a specific activity of 4.35 mCi/mmole, the total amount of free tyramine contained in the sample can be determined by dividing 50 μCi by 6.98×10^4 dpm/μg tyramine as follows.

$$\text{Total amount of tyramine (in } \mu\text{g)} = \frac{(50\ \mu\text{Ci})\ (2.2 \times 10^{6}\ \text{dpm}/\mu\text{Ci})}{6.98 \times 10^{4}\ \text{dpm}/\mu\text{g}}$$

$$= 1575\ \mu\text{g}$$

Because the solution is to contain 50 μg tyramine per milliliter, 1575 μg divided by 50 gives 31.5 ml, which is the final volume to which the 50 μCi of radioactive tyramine must be diluted.

It should be pointed out that this is a fairly active solution. Each 0.1-ml aliquot of this solution contains a total of 349,000 dpm. If this solution is to be used in a routine determination of monoamine oxidase activity in biological specimens, it is desirable to make the solution to a predetermined strength in terms of radioactive content. An adequate level for most applications employing ^{14}C is 100,000 dpm per analysis. This amount of radioactivity gives 50,000 cpm at 50% counting efficiency. This means that as little as 1% utilization of substrate with this starting amount of radioactivity yields a radioactive metabolite containing 500 cpm. A sample containing 500 cpm is very easy to measure in a scintillation counter. A 10-minute count of this sample results in an accumulation of 5000 net counts. The counting error (standard deviation) for this number of counts is less than 2%.

If we now desire to make up the radioactive tyramine solution so that each 0.1-ml aliquot contained 5 μg of tyramine with a total radioactivity of 100,000 dpm from the same tyramine hydrobromide used in the previous example, the procedure is as follows.

Because we have started with 50 μCi of tyramine hydrobromide, we have a total of (50 μCi) (2.2×10^6 dpm/μCi) or 1.1×10^8 dpm. Each milliliter of the desired radioactive tyramine should contain exactly 1,000,000 dpm or 1×10^6 dpm/ml. Therefore, 50 μCi of tyramine must be diluted to

$$\frac{1.1 \times 10^8 \text{ dpm}}{1 \times 10^6 \text{ dpm/ml}} = 110 \text{ ml}$$

Because the final concentration of tyramine must be 50 μg/ml, there must be a total of (110 ml) (50 μg tyramine/ml) or 5500 μg of tyramine contained in this solution.

From the specific radioactivity of the tyramine hydrobromide given as 4.35 mCi/mmole, we know that the total amount of free tyramine contained in 50 μCi is

$$\frac{50\ \mu\text{Ci}}{4.35\ \mu\text{Ci}/\mu\text{mole}}\ 137\ \mu\text{g}/\mu\text{mole} = 1575\ \mu\text{g tyramine}$$

Because the required amount of tyramine is 5500 μg and the 50 μCi of radioactive tyramine contains 1575 μg of tyramine, we must add the difference as nonisotopic tyramine.

$$5500 - 1575 = 3925 \text{ or } 3.9 \text{ mg tyramine, free base}$$

Nonisotopic tyramine is available as tyramine hydrochloride instead of as the hydrobromide. Therefore, the amount of nonisotopic tyramine hydrochloride equivalent to 3.9 mg of tyramine is

$$(3.9 \text{ mg tyramine}) \frac{(173.5 \text{ mg tyramine HCl})}{(137 \text{ mg tyramine})} = 4.94 \text{ mg tyramine HCl}$$

In summary, in order to make a [^{14}C] tyramine solution containing 50 μg tyramine (free base) per milliliter with an activity of 1 million dpm/ml, you dissolve the radioactive tyramine, specific activity 4.35 mCi/mmole, and 4.94 mg of nonisotopic tyramine hydrochloride in a final volume of 110 ml. This gives the desired solution. One-tenth milliliter of this solution contains 5 μg of [^{14}C]tyramine (free base) with a total radioactive content of 100,000 dpm.

DECONTAMINATION OF GLASS COUNTING VIALS AND RADIOACTIVE GLASSWARE

One of the major expenses of sample preparation for liquid scintillation counting is the cost of the vials. Where critical work is done, it is essential to have uniform and low background, and in these instances, washing and reusing vials is not advocated. However, for screening purposes, such as counting sequential samples being eluted from a column or for assaying samples containing more than a few hundred decompositions per minute of activity, washed vials are adequate for the task.

Kushinsky and Paul (1) have reported on a radioactive glassware decontamination procedure that they have used successfully for several years. It entails an initial rinse of the glassware in solvent mixture of acetone–methanol–benzene, 2:1:1 (v/v), followed by a 2-hour wash in a Rethxo washer using benzene–methanol, 1:1, as a solvent. These vials are then dried in a self-cleaning oven for 3–8 hours at about 470°C and the oven is allowed to cool 2 hours before opening the door. These authors have washed over 10,000 vials with no obvious evidence of failure of the decontamination process. The oven used was an ordinary, inexpensive home appliance with a self-cleaning feature.

It appears reasonable that vials that are initially rinsed free of the counting mixture with acetone, followed by an aqueous detergent wash, can be treated in a self-cleaning oven to eliminate the radioactive contaminants by heat and oxidation.

PRECISION OF FIXED-TIME VERSUS FIXED-COUNT MEASUREMENTS

The question of the statistical advantage of counting samples for a fixed time or to a predetermined number of counts is probably one of the most commonly asked in radioisotopic analysis. This question has been considered by Gaylor (2), who has reasoned that a fair comparison can be made mathematically if the problem is analyzed by setting the total counting time for the sample plus background equal for the two techniques. When this is done and the arithmetic computed, the conclusion reached is that the variance of the fixed-time method is always less than or equal to the variance of the fixed-count method for any given total counting time. Conversely, for a specified precision, the expected total counting time required for the fixed-time method is less than or equal to the expected total counting time required for the fixed-count method.

REFERENCES

1. Kushinsky, S., and Paul, W. *Anal. Biochem.* **30,** 468 (1969).
2. Gaylor, D. W. *Anal. Chem.* **34,** 1670 (1962).

ISOTOPE	PARTICLE TYPE	ENERGY MeV	HALF-LIFE
^{3}H	β^-	0.018	12.3 YEARS
^{14}C	β^-	0.155	5568 YEARS
^{22}Na	β^+ γ γ†	0.54 1.277 0.51	2.58 YEARS
^{24}Na	β^- γ	1.39 1.37, 2.75	14.97 HOURS
^{32}P	β^-	1.71	14.22 DAYS
^{35}S	β^-	0.167	86.7 DAYS
^{36}Cl	β^-	0.714	3.08×10^{5} YEARS
^{40}K	β^- γ X-RAYS	1.33 1.46 0.0032	1.25×10^{9} YEARS
^{45}Ca	β^-	0.254	164 DAYS
^{51}Cr	γ X-RAYS	0.325 0.0055	27.8 DAYS
^{55}Fe	X-RAYS	0.0065	2.60 YEARS
^{59}Fe	β^- γ	0.27, 0.46 1.1, 1.3	45.1 DAYS
^{63}Ni	β^-	0.067	125 YEARS
^{65}Zn	β^+(1.5% OF DECAYS) γ X-RAYS	0.324 1.12 0.009	245 DAYS
^{90}Sr	β^-	0.545	27.7 YEARS
^{125}I	γ X-RAYS	0.035 0.033	60 DAYS
^{129}I	β^- γ	0.15 0.038	1.7×10^{7} YEARS
^{131}I	β γ	0.335, 0.608 0.284, 0.364 0.637	8.08 DAYS
^{137}Cs/^{137}Ba	β	0.514, 1.20 0.66	26.6 YEARS

†ANNIHILATION RADIATION

Fig. 59 Some of the radionuclides that can be assayed by liquid scintillation counting.

YEARS	MONTHS					
	0	2	4	6	8	10
0	1.00	.99	.98	.97	.96	.95
1	.95	.94	.93	.92	.91	.90
2	.89	.88	.88	.87	.86	.85
3	.84	.84	.83	.82	.81	.80
4	.80	.79	.78	.78	.77	.76
5	.75	.75	.74	.73	.73	.72
6	.71	.71	.70	.69	.69	.68
7	.67	.67	.66	.65	.65	.64
8	.64	.63	.62	.62	.61	.61
9	.60	.60	.59	.58	.58	.57
10	.57	.56	.56	.55	.55	.54
11	.54	.53	.53	.52	.52	.51
12	.51	.50				

Fig. 60 Tritium decay table. $T_{1/2} = 12.26$ years.

DAYS	DAYS				
	0	2	4	6	8
0	1.0000	.9841	.9685	.9532	.9380
10	.9232	.9085	.8941	.8799	.8660
20	.8523	.8387	.8254	.8124	.7995
30	.7868	.7743	.7620	.7500	.7381
40	.7264	.7148	.7035	.6923	.6814
50	.6706	.6599	.6495	.6392	.6290
60	.6190	.6092	.5996	.5901	.5807
70	.5715	.5624	.5535	.5447	.5361
80	.5276	.5192	.5110	.5029	.4949
90	.4871	.4793	.4717	.4642	.4569
100	.4496	.4425	.4355	.4286	.4218
110	.4151	.4085	.4020	.3957	.3894
120	.3832	.3771	.3712	.3653	.3595
130	.3538	.3482	.3426	.3372	.3319
140	.3266	.3214	.3163	.3113	.3064
150	.3015	.2967	.2920	.2874	.2828
160	.2783	.2739	.2696	.2653	.2611
170	.2570	.2529	.2489	.2449	.2410
180	.2372	.2335	.2298	.2261	.2225
190	.2190	.2155	.2121	.2087	.2054
200	.2022	.1990	.1958	.1927	.1897

Fig. 61 Sulfur-35 decay table. $T_{1/2} = 86.7$ days.

DAYS ELAPSED	FRACTION REMAINING	DAYS ELAPSED	FRACTION REMAINING	DAYS ELAPSED	FRACTION REMAINING
1	0.95265	41	0.13687	81	0.01966
2	0.90755	42	0.13039	82	0.01873
3	0.86457	43	0.12421	83	0.01785
4	0.82364	44	0.11833	84	0.01700
5	0.78464	45	0.11273	85	0.01620
6	0.74749	46	0.10739	86	0.01543
7	0.71210	47	0.10231	87	0.01470
8	0.67838	48	0.09746	88	0.01400
9	0.64626	49	0.09285	89	0.01334
10	0.61565	50	0.08845	90	0.01271
11	0.58651	51	0.08426	91	0.01211
12	0.55874	52	0.08027	92	0.01153
13	0.53229	53	0.07647	93	0.01099
14	0.50708	54	0.07285	94	0.01047
15	0.48307	55	0.06940	95	0.00997
16	0.46020	56	0.06612	96	0.00950
17	0.43841	57	0.06299	97	0.00905
18	0.41765	58	0.06000	98	0.00862
19	0.39788	59	0.05716	99	0.00821
20	0.37904	60	0.05446	100	0.00782
21	0.36109	61	0.05188	101	0.00745
22	0.34400	62	0.04942	102	0.00710
23	0.32771	63	0.04708	103	0.00676
24	0.31219	64	0.04485	104	0.00644
25	0.29741	65	0.04273	105	0.00614
26	0.28333	66	0.04071	106	0.00585
27	0.26991	67	0.03878	107	0.00557
28	0.25713	68	0.03694	108	0.00531
29	0.24496	69	0.03519	109	0.00506
30	0.23336	70	0.03353	110	0.00482
31	0.22231	71	0.03194	111	0.00459
32	0.21178	72	0.03043	112	0.00437
33	0.20176	73	0.02899	113	0.00416
34	0.19220	74	0.02761	114	0.00397
35	0.18310	75	0.02631	115	0.00378
36	0.17443	76	0.02506	116	0.00360
37	0.16618	77	0.02387	117	0.00343
38	0.15831	78	0.02274	118	0.00327
39	0.15081	79	0.02167	119	0.00311
40	0.14367	80	0.02064	120	0.00297

Fig. 62 Phosphorus-32 decay table. $T_{1/2} = 14.29$ days. After I. W. Goodier and D. H. Pritchard, *Int. J. Appl. Rad. Isotopes* **17**, 121 (1966).

DAYS	DAYS									
	0	1	2	3	4	5	6	7	8	9
0	1.000	.9885	.9772	.9659	.9549	.9439	.9330	.9223	.9117	.9013
10	.8909	.8807	.8706	.8606	.8507	.8409	.8313	.8217	.8123	.8029
20	.7937	.7846	.7756	.7667	.7579	.7492	.7406	.7320	.7236	.7153
30	.7071	.6990	.6910	.6830	.6752	.6674	.6597	.6522	.6447	.6372
40	.6299	.6227	.6155	.6085	.6015	.5946	.5877	.5810	.5743	.5677
50	.5612	.5548	.5484	.5421	.5359	.5297	.5236	.5176	.5116	.5057
60	.5000	.4943	.4886	.4830	.4774	.4719	.4665	.4612	.4559	.4506
70	.4454	.4403	.4353	.4303	.4253	.4204	.4156	.4108	.4061	.4014
80	.3968	.3923	.3877	.3833	.3789	.3745	.3702	.3660	.3618	.3576
90	.3535	.3494	.3454	.3414	.3375	.3336	.3298	.3260	.3223	.3186
100	.3149	.3113	.3077	.3042	.3007	.2972	.2938	.2904	.2871	.2838
110	.2805	.2773	.2741	.2709	.2678	.2647	.2617	.2587	.2557	.2528
120	.2500	.2471	.2443	.2415	.2387	.2360	.2332	.2306	.2279	.2253
130	.2227	.2201	.2176	.2151	.2126	.2102	.2078	.2054	.2030	.2007
140	.1984	.1961	.1938	.1916	.1894	.1872	.1851	.1829	.1808	.1788
150	.1767	.1747	.1727	.1707	.1687	.1668	.1648	.1629	.1611	.1592
160	.1574	.1556	.1538	.1520	.1503	.1485	.1468	.1451	.1435	.1418
170	.1402	.1386	.1370	.1354	.1338	.1323	.1308	.1293	.1278	.1263
180	.1250	.1236	.1221	.1207	.1193	.1180	.1166	.1153	.1139	.1126

Fig. 63 Iodine-125 decay table. $T_{1/2}$ = 60 days.

DAYS	HOURS											
	0	2	4	6	8	10	12	14	16	18	20	22
0	1.0000	.9929	.9858	.9788	.9718	.9649	.9580	.9512	.9444	.9377	.9310	.9244
1	.9178	.9112	.9047	.8983	.8919	.8856	.8792	.8730	.8668	.8606	.8545	.8484
2	.8423	.8363	.8304	.8244	.8186	.8128	.8070	.8012	.7955	.7898	.7842	.7786
3	.7731	.7676	.7621	.7567	.7513	.7459	.7406	.7353	.7301	.7249	.7197	.7146
4	.7095	.7045	.6995	.6945	.6895	.6846	.6797	.6749	.6701	.6653	.6606	.6559
5	.6512	.6466	.6419	.6374	.6328	.6283	.6238	.6194	.6150	.6106	.6063	.6019
6	.5977	.5934	.5892	.5850	.5808	.5767	.5726	.5685	.5644	.5604	.5564	.5525
7	.5485	.5446	.5407	.5369	.5331	.5293	.5255	.5217	.5180	.5143	.5106	.5070
8	.5034	.4998	.4963	.4927	.4892	.4857	.4823	.4789	.4754	.4721	.4687	.4654
9	.4620	.4587	.4555	.4522	.4490	.4458	.4426	.4395	.4364	.4332	.4302	.4271
10	.4240	.4210	.4180	.4151	.4121	.4092	.4063	.4034	.4005	.3976	.3948	.3920
11	.3892	.3864	.3837	.3809	.3782	.3755	.3728	.3702	.3676	.3649	.3623	.3598
12	.3572	.3546	.3521	.3496	.3471	.3446	.3422	.3398	.3373	.3349	.3325	.3302
13	.3278	.3255	.3232	.3209	.3186	.3163	.3141	.3118	.3096	.3074	.3052	.3030
14	.3009	.2987	.2966	.2945	.2924	.2903	.2882	.2862	.2842	.2821	.2801	.2781
15	.2761	.2742	.2722	.2703	.2683	.2664	.2645	.2627	.2608	.2589	.2571	.2553
16	.2534	.2516	.2498	.2481	.2463	.2445	.2428	.2411	.2394	.2376	.2360	.2343
17	.2326	.2309	.2293	.2277	.2261	.2245	.2228	.2212	.2196	.2181	.2165	.2150
18	.2135	.2120	.2105	.2090	.2075	.2060	.2045	.2030	.2116	.2002	.1987	.1973
19	.1959	.1945	.1931	.1918	.1904	.1890	.1877	.1864	.1851	.1837	.1824	.1811
20	.1798	.1785	.1772	.1760	.1747	.1735	.1723	.1711	.1699	.1686	.1674	.1662
21	.1650	.1638	.1626	.1615	.1603	.1692	.1581	.1570	.1559	.1548	.1537	.1526
22	.1515	.1504	.1493	.1482	.1471	.1461	.1451	.1441	.1430	.1420	.1410	.1400
23	.139	.138	.137	.136	.135	.134	.133	.132	.131	.130	.129	.128
24	.127	--	--	.125								

Fig. 64 Iodine-131 decay table. $T_{1/2}$ = 8.08 days.

The following nomographs have been compiled by A. A. Jarrett, "Statistical Methods in the Measurement of Radioactivity with Some Useful Graphs and Nomographs," AECU-262. U. S. At. Energy Comm., Oak Ridge, Tennessee, 1946.

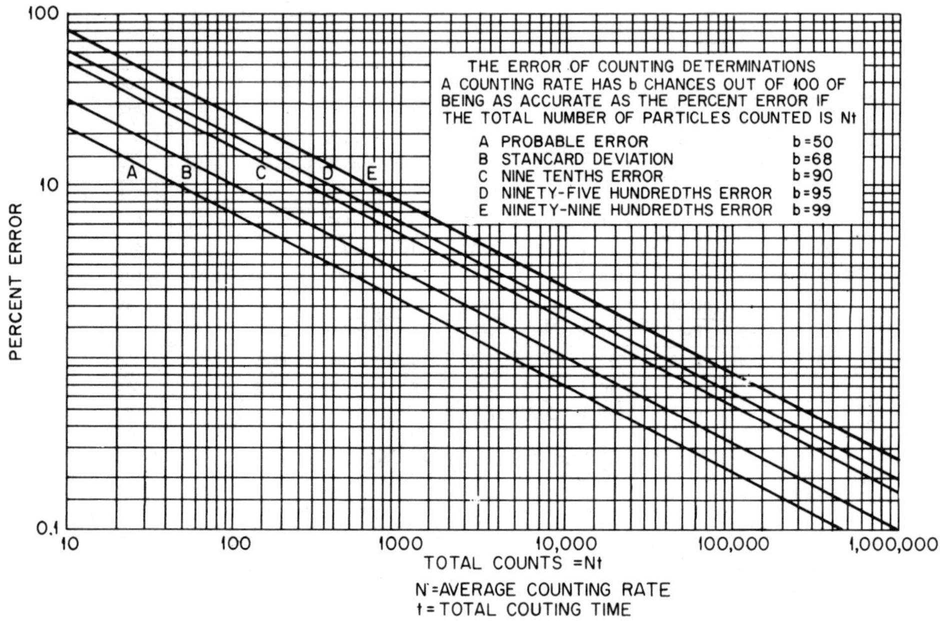

Fig. 65 The error of counting determinations.

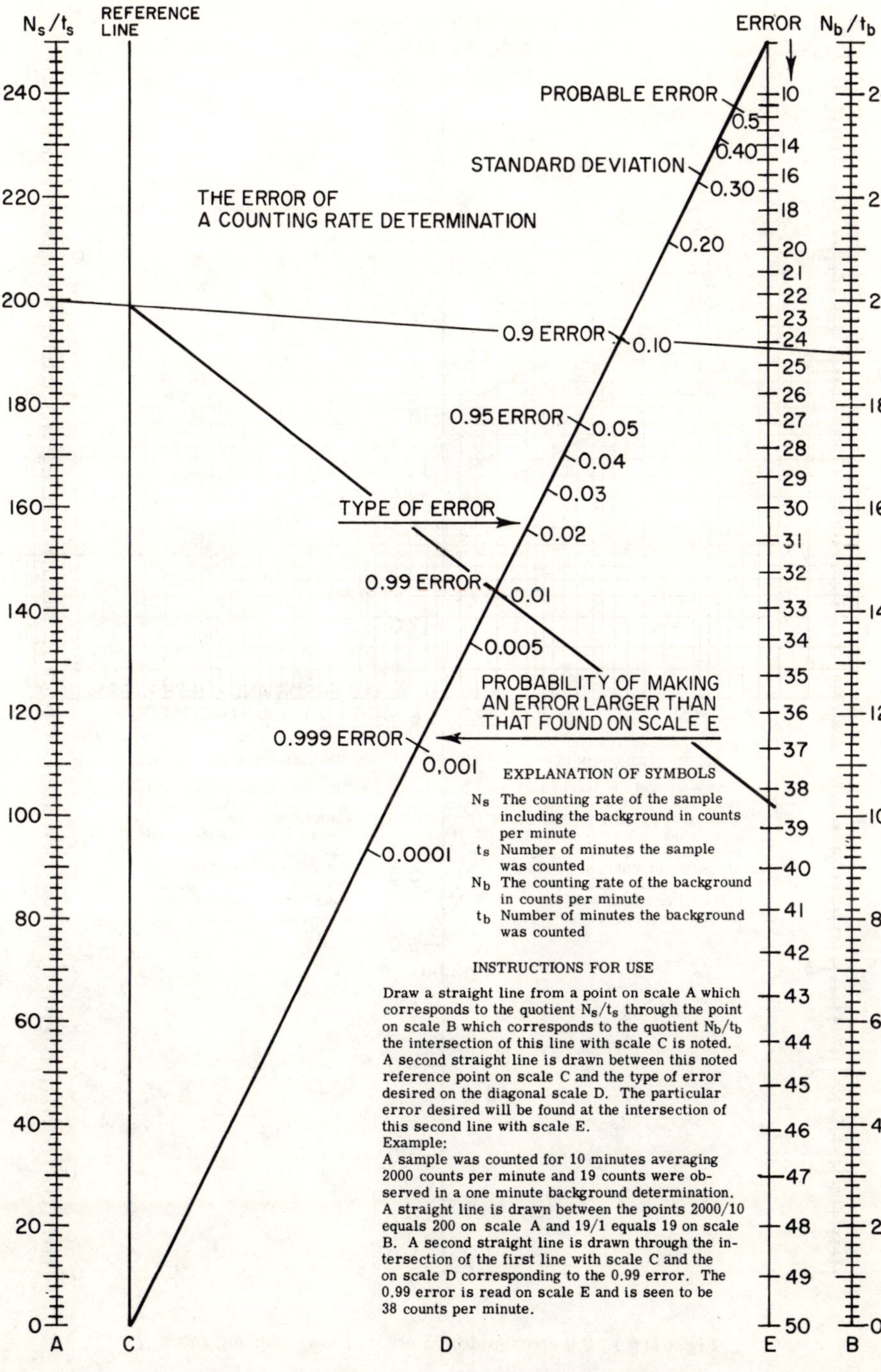

Fig. 66 The error of a counting rate determination.

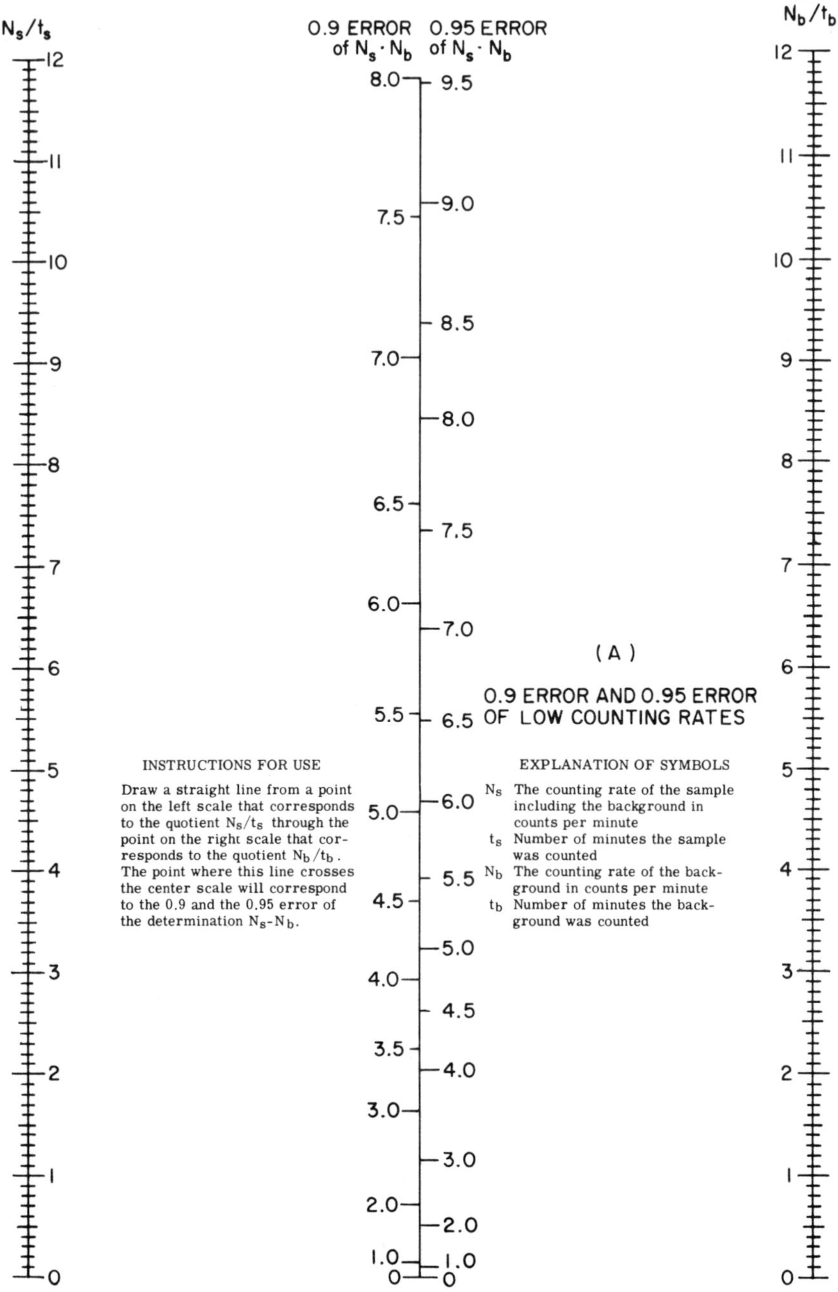

Fig. 67 (A) 0.9 error and 0.95 error of low counting rates.

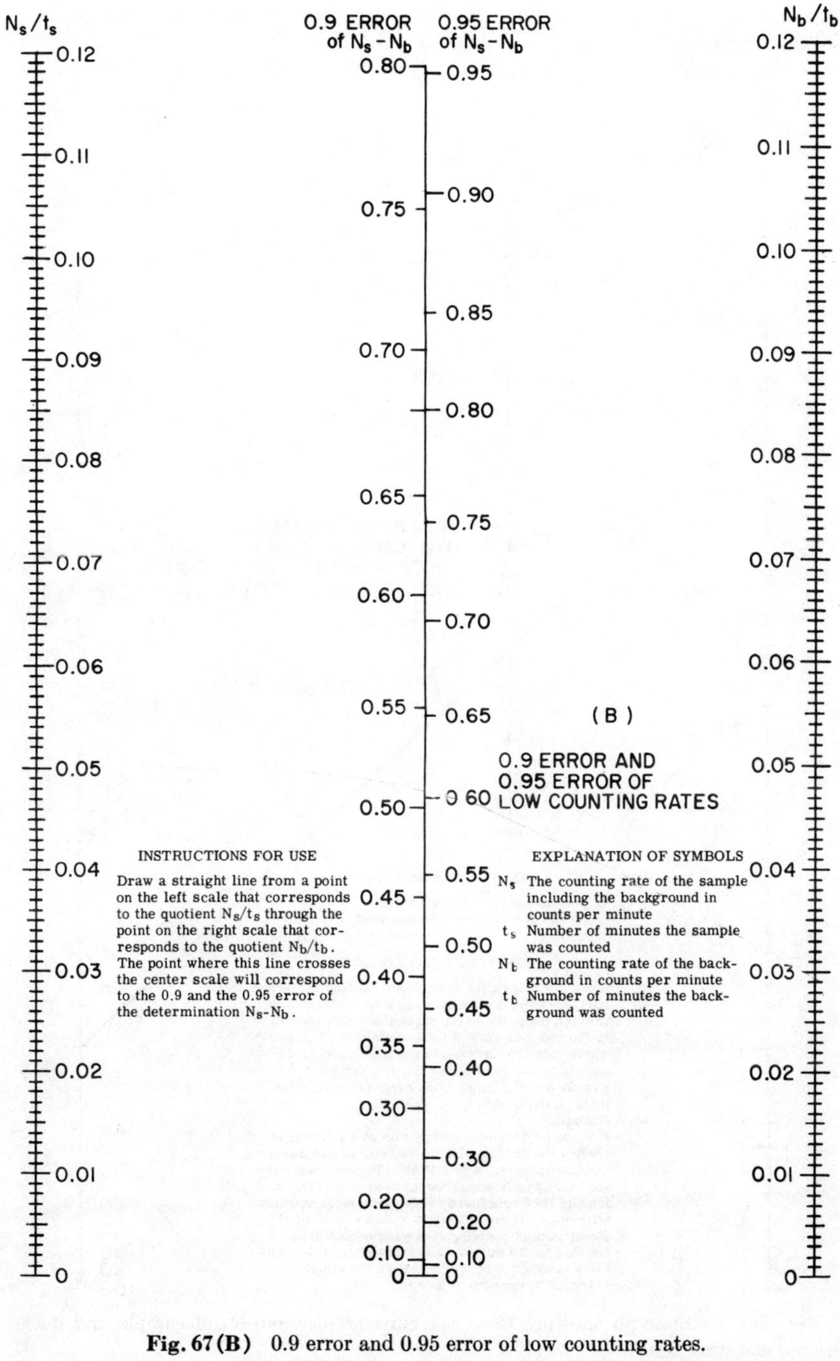

Fig. 67 (B) 0.9 error and 0.95 error of low counting rates.

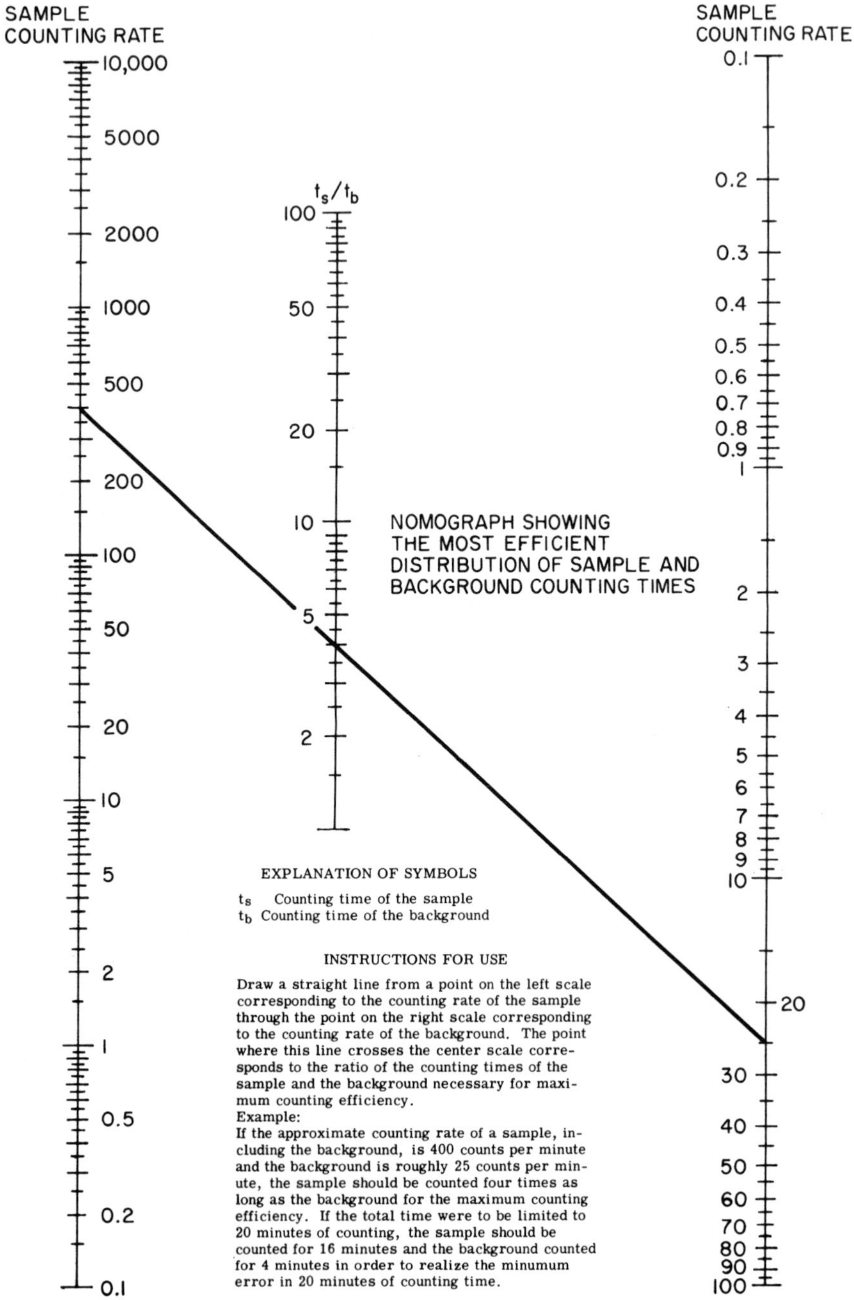

Fig. 68 Nomograph showing the most efficient distribution of sample and background counting times.

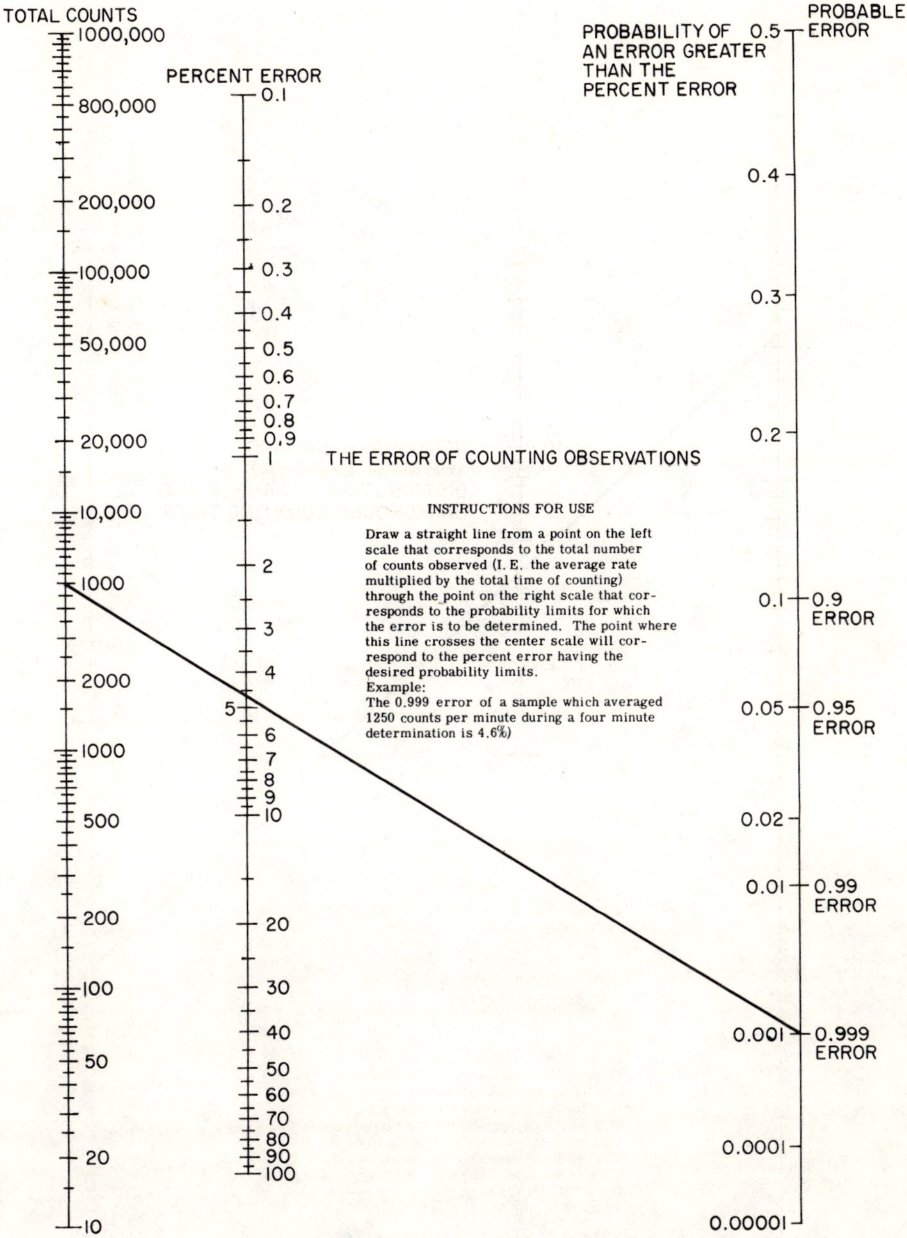

Fig. 69 Nomograph showing the error of counting determinations.

Index

D

M

N

O

P

Q

R

A 4
B 5
C 6
D 7
E 8
F 9
G 0
H 1
I 2
J 3